Communications
in Computer and Information Science   737

Commenced Publication in 2007
Founding and Former Series Editors:
Alfredo Cuzzocrea, Dominik Ślęzak, and Xiaokang Yang

More information about this series at http://www.springer.com/series/7899

Chiara Francalanci · Markus Helfert (Eds.)

Data Management Technologies and Applications

5th International Conference, DATA 2016
Colmar, France, July 24–26, 2016
Revised Selected Papers

 Springer

Editors
Chiara Francalanci
Department of Electronics and Information
Politecnico di Milano
Milan
Italy

Markus Helfert
School of Computing
Dublin City University
Dublin
Ireland

ISSN 1865-0929 ISSN 1865-0937 (electronic)
Communications in Computer and Information Science
ISBN 978-3-319-62910-0 ISBN 978-3-319-62911-7 (eBook)
DOI 10.1007/978-3-319-62911-7

Library of Congress Control Number: 2017945731

Printed on acid-free paper

This Springer imprint is published by Springer Nature
The registered company is Springer International Publishing AG
The registered company address is: Gewerbestrasse 11, 6330 Cham, Switzerland

Preface

The present book includes extended and revised versions of a set of selected papers from the 5th International Conference on Data Management Technologies and Applications (DATA 2016), held in Lisbon, Portugal, during July 24–26, 2016.

We received 50 paper submissions from 26 countries, of which 18% are included in this book. The papers were selected by the event chairs and their selection is based on a number of criteria that include the classifications and comments provided by the Program Committee members, the session chairs' assessment, and also the program chairs' global view of all papers included in the technical program. The authors of selected papers were then invited to submit a revised and extended version of their papers having at least 30% innovative material.

The 5th International Conference on Data Management Technologies and Applications (DATA) aims to bring together researchers, engineers and practitioners interested in databases, data warehousing, data mining, data management, data security and other aspects of information systems and technology involving advanced applications of data.

The papers selected to be included in this book contribute to the understanding of relevant trends of current research on business analytics, data management and quality, ontologies and the Semantic Web, and databases and data security. They substantially contribute to the literature by providing interesting use cases, demonstrating the application of modern big data design and management techniques. The interdisciplinary approach of the DATA conference is a fundamental enabler of this scientific contribution. Particularly, the ability to attract both academics and practitioners and a focus on technology application represent important drivers of the practical consequence of the papers published in this book. The papers address a range of specific topics including: data consistency on agnostic fault-tolerant systems, improving performances of an embedded relational database management system, visual citation tracing, efficient multi-domain data processing, pay-as-you-go data quality management, real-time analytics of streaming data, computation of belief combination rules, identifying conversational message threads, and ETL processes specification using a pattern-based ontology. Overall, they provide an overview of the state of the art, with a focus on data and on their impact on business processes. Organizational policies, technical choices, and economic benefits from the exploitation of modern technologies are largely addressed by the papers published in this book, with a broad set of references to the most useful literature on the subject.

We would like to thank all the authors for their contributions and also the reviewers who helped ensure the quality of this publication.

February 2017

Chiara Francalanci
Markus Helfert

Organization

Conference Chair

Markus Helfert   Dublin City University, Ireland

Program Chair

Chiara Francalanci   Politecnico di Milano, Italy

Program Committee

Muhammad Abulaish   South Asian University, India
Hamideh Afsarmanesh   University of Amsterdam, The Netherlands
Markus Aleksy   ABB Corporate Research Center, Germany
Christos Anagnostopoulos University of Glasgow, UK
Nicolas Anciaux   Inria Paris-Rocquencourt, France
Kenneth Anderson   University of Colorado, USA
Keijiro Araki   Kyushu University, Japan
Bernhard Bauer   University of Augsburg, Germany
Andreas Behrend   University of Bonn, Germany
Fevzi Belli   Izmir Institute of Technology, Turkey
Karim Benouaret   Université Claude Bernard Lyon 1, France
Jorge Bernardino   Polytechnic Institute of Coimbra - ISEC, Portugal
Francesco Buccafurri   University of Reggio Calabria, Italy
Dumitru Burdescu   University of Craiova, Romania
Kung Chen   National Chengchi University, Taiwan
Yangjun Chen   University of Winnipeg, Canada
Byron Choi   Hong Kong Baptist University, Hong Kong,
                            SAR China
Christine Collet   Grenoble Institute of Technology, France
Agostino Cortesi   Università Ca' Foscari di Venezia, Italy
Theodore Dalamagas   Athena Research and Innovation Center, Greece
Bruno Defude   Institut Mines Telecom, France
Steven Demurjian   University of Connecticut, USA
Stefan Dessloch   Kaiserslautern University of Technology, Germany
Fabien Duchateau   Université Claude Bernard Lyon 1/LIRIS, France
Todd Eavis   Concordia University, Canada
Mohamed Y. Eltabakh   Worcester Polytechnic Institute, USA
Markus Endres   University of Augsburg, Germany
Sergio Firmenich   Universidad Nacional de La Plata, Argentina
Kehan Gao   Eastern Connecticut State University, USA
Roberto García   Universitat de Lleida, Spain

Jérôme Gensel	Université Grenoble Alpes, France
Paola Giannini	University of Piemonte Orientale, Italy
Giorgos Giannopoulos	Athena Research and Innovation Center, Greece
J. Paul Gibson	Mines-Telecom, Telecom SudParis, France
Matteo Golfarelli	University of Bologna, Italy
Janis Grabis	Riga Technical University, Latvia
Jerzy Grzymala-Busse	University of Kansas, USA
Mena Habib	Maastricht University, The Netherlands
Raju Halder	Indian Institute of Technology Patna, India
Waqar Haque	University of Northern British Columbia, Canada
Andreas Henrich	University of Bamberg, Germany
Jose Luis Arciniegas Herrera	Universidad del Cauca, Colombia
Jang-Eui Hong	Chungbuk National University, Korea, Republic of
Tsan-Sheng Hsu	Institute of Information Science, Academia Sinica, Taiwan
Ivan Ivanov	SUNY Empire State College, USA
Wang Jianmin	Tsinghua University, China
Konstantinos Kalpakis	University of Maryland Baltimore County, USA
Dimitris Karagiannis	University of Vienna, Austria
Maurice van Keulen	University of Twente, The Netherlands
Benjamin Klöpper	ABB Corporate Research, Germany
Mieczyslaw Kokar	Northeastern University, USA
Kostas Kolomvatsos	University of Thessaly, Greece
John Krogstie	NTNU, Norway
Martin Krulis	Charles University, Czech Republic
Konstantin Läufer	Loyola University Chicago, USA
Dominique Laurent	ETIS Laboratory CNRS UMR 8051, Cergy-Pontoise University, ENSEA, France
Sangkyun Lee	TU Dortmund, Germany
Raimondas Lencevicius	Nuance Communications, USA
Haikun Liu	Huazhong University of Science and Technology, China
Ricardo J. Machado	Universidade do Minho, Portugal
Zaki Malik	Wayne State University, USA
Keith Marsolo	Cincinnati Children's Hospital Medical Center, USA
Miguel A. Martínez-Prieto	University of Valladolid, Spain
Florent Masseglia	Inria, France
Fabio Mercorio	University of Milano-Bicocca, Italy
Dimitris Mitrakos	Aristotle University of Thessaloniki, Greece
Stefano Montanelli	Università degli Studi di Milano, Italy
Bongki Moon	Seoul National University, Korea, Republic of
Gianluca Moro	Università di Bologna, Italy
Mirella M. Moro	Federal University of Minas Gerais (UFMG), Brazil
Mikhail Moshkov	KAUST, Saudi Arabia
Josiane Mothe	Université de Toulouse, France

Dariusz Mrozek	Silesian University of Technology, Poland
Richi Nayak	Queensland University of Technology, Australia
Paulo Novais	Universidade do Minho, Portugal
Boris Novikov	Saint Petersburg University, Russian Federation
Jisha Jose Panackal	Vidya Academy of Science and Technology, India
George Papastefanatos	Athena Research and Innovation Center, Greece
Jeffrey Parsons	Memorial University of Newfoundland, Canada
Barbara Pernici	Politecnico di Milano, Italy
Ilia Petrov	Reutlingen University, Germany
Iulian Sandu Popa	University of Versailles Saint-Quentin-en-Yvelines and Inria Saclay, France
Nirvana Popescu	University Politehnica of Bucharest, Romania
Philippe Pucheral	University of Versailles Saint-Quentin en Yvelines (UVSQ), France
Elisa Quintarelli	Politecnico di Milano, Italy
Christoph Quix	RWTH Aachen University, Germany
Praveen Rao	University of Missouri-Kansas City, USA
Alexander Rasin	DePaul University, USA
Kun Ren	Yale University, USA
Colette Rolland	Université De Paris1 Panthèon Sorbonne, France
Gustavo Rossi	Lifia, Argentina
Gunter Saake	Institute of Technical and Business Information Systems, Germany
Dimitris Sacharidis	Technische Universität Wien, Austria
Manuel Filipe Santos	Centro ALGORITMI, University of Minho, Portugal
Maria Luisa Sapino	Università di Torino, Italy
Ralf Schenkel	University of Trier, Germany
Diego Seco	University of Concepción, Chile
Vinay Setty	Max Planck Institut für Informatik, Germany
lijun shan	Inria, France
Nematollaah Shiri	Concordia University, Canada
Harvey Siy	University of Nebraska at Omaha, USA
Spiros Skiadopoulos	University of Peloponnese, Greece
Yeong-Tae Song	Towson University, USA
Sergey Stupnikov	IPI RAN, Russian Federation
Zbigniew Suraj	University of Rzeszow, Poland
Neamat El Tazi	Cairo University, Egypt
Maguelonne Teisseire	Irstea, National Research Institute of Science and Technology for Environment and Agriculture, France
Manolis Terrovitis	Institute for the Management of Information Systems, Greece
Babis Theodoulidis	University of Manchester, UK
Frank Tompa	University of Waterloo, Canada
Christos Tryfonopoulos	University of Peloponnese, Greece

Goran Velinov	UKIM, Macedonia, Former Yugoslav Republic of
Thanasis Vergoulis	Athena Research and Innovation Center, Greece
Karin Verspoor	University of Melbourne, Australia
José Ríos Viqueira	Universidade de Santiago de Compostela, Spain
Gianluigi Viscusi	EPFL Lausanne, Switzerland
Hannes Voigt	TU Dresden, Germany
Florian Wenzel	University of Augsburg, Germany
Leandro Krug Wives	Universidade Federal do Rio Grande do Sul, Brazil
Robert Wrembel	Poznan University of Technology, Poland
Yun Xiong	Fudan University, China
Filip Zavoral	Charles University Prague, Czech Republic
Jiakui Zhao	State Grid Information and Telecommunication Group of China, China
Jianlong Zhong	GRAPHSQL Inc., USA
Yangyong Zhu	Fudan University, China

Additional Reviewers

Idir Benouaret	Heudiasyc Laboratory, UMR CNRS 7253, Université de Technologie de Compiègne, France
Dominik Bork	University of Vienna, Austria
Michele A. Brandao	Universidade Federal de Minas Gerais, Brazil
Estrela Ferreira Cruz	Instituto Politécnico de Viana do Castelo, Portugal
Giacomo Domeniconi	University of Bologna, Italy
José Fuentes	University of Concepción, Chile
Olga Gkountouna	Research Center Athena, Greece
Emanuele Rabosio	Politecnico di Milano, Italy

Invited Speakers

| Panos Vassiliadis | University of Ioannina, Greece |
| Christoph Quix | RWTH Aachen University, Germany |

Contents

A Scalable Platform for Low-Latency Real-Time Analytics of Streaming Data

Paolo Cappellari[1]([✉]), Mark Roantree[2], and Soon Ae Chun[1]

[1] City University of New York, New York, USA
{paolo.cappellari,soon.chun}@csi.cuny.edu
[2] School of Computing, Insight Centre for Data Analytics,
Dublin City University, Dublin, Ireland
mark.roantree@cs.dcu.ie

Abstract. The ability to process high-volume high-speed streaming data from different data sources is critical for modern organizations to gain insights for business decisions. In this research, we present the streaming analytics platform (SDAP), which provides a set of operators to specify the process of stream data transformations and analytics. SDAP adopts a declarative approach to model and design, delivering analytics capabilities through the combination of a set of primitive operators in a simple manner. The model includes a topology to design streaming analytics specifications using a set of atomic data manipulation operators. Our evaluation demonstrates that SDAP is capable of maintaining low-latency while scaling to a cloud of distributed computing nodes, and providing easier process design and execution of streaming analytics.

Keywords: Data stream processing · High-performance computing · Low-latency · Distributed systems

1 Introduction

In their quest for competitive advantage, extending data analysis to include streaming data sources has become a requirement for the majority of organizations. Driven by the need for more timely results and having to deal with an increasing availability of real-time data sources, companies are investing in integrating data streaming processing systems in their applications stack. Real-time data processing can help multiple application domains, such as stock trading, new product monitoring, fraud detection and regulatory compliance monitoring, supporting situation awareness and decision making with real-time alerts and real-time analytics. Real-time data processing and analytics requires flexible integration of live data captured from different sources that would otherwise be lost, with traditional data from enterprise storage repositories (e. g. data warehouse).

Existing streaming systems mainly focus on the problems of scalability, fault-tolerance, flexibility, and performance of individual operations, e.g.

© Springer International Publishing AG 2017
C. Francalanci and M. Helfert (Eds.): DATA 2016, CCIS 737, pp. 1–24, 2017.
DOI: 10.1007/978-3-319-62911-7_1

[1,2,26,29]. Our approach to building a high-performance data stream processing engine was influenced by the growing need of more timely information by organizations and the success of streaming systems such as Yahoo S4 [22], Storm [27], Sonora [6], and Spark-Streaming [29]. We observed that none of the modern systems target low-latency and high-performance while also providing an easy way of developing streaming applications for non-expert users. Unlike these systems and other related research, we focus on the provision of a complete and comprehensive solution for the rapid development, execution and management of scalable high-performance, low-latency, stream analytics applications.

1.1   Motivation and Case Study

To illustrate the complex tasks involved in a stream analytics process, we use a scenario which seeks to understand the performance of the bike utilization in multiple locations within a city, trying to monitor the trend of the performance data and comparing usage with bike usage in other cities. Assume a scenario where a town planner needs to know various performance indicators, such as whether bicycles are parked in specific docking stations located across the city are utilized to an acceptable level. This requires the constant monitoring of data from the Bike Sharing Systems (BSS). Every 60 s, the BSS reports the status of each station, which include the number of bikes docked at each station. The goal is to calculate the performance of the BSS as the number of bikes in utilization against the total number of bikes available in the system, in order to identify stations with lower than predicted usage or stations with high usage that require expansion. In addition, the manager is also interested in the performance of bike sharing program in other cities, to gain a direct comparison among different BSS.

1.2   Contribution

In this paper, we improve on our previous work [3] and we present the streaming data analytics platform (SDAP), which provides a set of operators to specify the process of stream data transformations and analytics, together with its execution environment. SDAP adopts a declarative approach to both modeling and designing a streaming analytics system using combinations of primitive operators in a straightforward manner.

SDAP is aimed to be a robust platform for flexible design of streaming analytics applications that addresses the following broad requirements:

- The processing engine can manage high volumes of streaming data even when the rate at which data generated is extremely high;
- Results of steaming analytics and processing, on which organizations base decisions, are available as soon as possible;
- It supports designers of analytical processes by abstracting from the underlying parallel computation or high-performance programming;
- It is easy to develop, maintain and optimize the analytical applications.

The contribution of this research can be summarized as follows:

- Streaming analytics model. SDAP provides a set of operators to support a declaration-based analytics development environment.
- Streaming analytics application specification. SDAP provides users without prior knowledge of parallel computation or high-performance. programming, the tools to easily specify a 'topology', which describes the analytics process.
- High-performance topology execution. SDAP delivers a platform that exploits the best performing hardware and software to execute a topology, while also efficiently managing the resource computation underlying data stores and parallel processing.

A comprehensive evaluation demonstrates the performance of our system, both in terms of latency and ease of development. SDAP presents the lowest latency among the compared systems with the same low latency maintained when scaling to a large number of computational resources. Compared to similar systems, SDAP is different because it provides each of the following characteristics: (i) it offers built-in operators optimized for parallel computation; (ii) it was designed to deliver the best latency performance by exploiting the high-performance hardware and software libraries; (iii) it is easy to use, since users are not required to have programming skills; and (iv) it enables rapid development, since applications are specified in a declarative way, where users link built-in operations in a pipeline fashion.

The paper is organized as follows. In Sect. 2, we provide a comparison of our approach against other works. Section 3 we demonstrate a use case realized by using SDAP, which is used as a running example throughout the paper. In Sect. 4, we define the modeling of our platform, including the constructs and the primitives. Section 5, discusses the platform's architecture. In Sect. 6, we discuss the experimental setting, the performance results, and the ease of usage compared to a popular alternative. Finally, in Sect. 7 we present our conclusions.

2   Related Research

Research projects such as S4 [22], IBM InfoSphere Streams [15], and Storm [27], are considered event-based streaming systems as they process each tuple as soon as they become available. While this is a requirement for low-latency system, these research projects do not address latency or high-performance directly. The S4 system, for instance, provides a programming model similar to Map-Reduce, where data is routed from one operation to the next on the basis of key values. In comparison to SDAP, their approach limits the ability of the designer in the development of generic streaming applications. Storm [27] offers a set of primitives to develop topologies. In brief, a set of constructs are provided to route data between operations, similar to our approach in SDAP. However, the developer must provide the implementation at each step in the topology and thus, requiring development effort and expertise in parallel programming. This is not the case in SDAP, where built-in operations are provided, so that designers can focus on the creation of the topology rather than on the *implementation* of the operators. IBM InfoSphere Streams [15] follows an approach similar to Storm but also offers a set of predefined operations. In fact, designers can assemble operations in workflows, very much like in SDAP. However, the InfoSphere Streams approach puts the focus on quality of service of the topologies, rather than on latency performance.

A system that specifically targets low latency stream processing is Google's MillWheel [1]. As with the systems described above, MillWheel adopts an event based design and in this case, data manipulation operations are specified in a topology fashion. Similarly to S4, the computation paradigm is based on a key model: data is routed between computation resources on the basis of the value a key holds in the data. As it is the case for S4, this paradigm facilitates the evaluation of operations requiring grouping (on the same key), and only guarantee pure distributed parallelism using different keys. SDAP offers greater flexibility in this respect as the SDAP designer can choose whether or not to base the computation on keys. Moreover, as shown in our evaluation, SDAP delivers superior performance.

In an entirely different approach, the research ideas presented in [7,11,23, 28,29], approach data stream processing by embracing a micro-batch oriented design. These approaches extend the Map-Reduce paradigm and Hadoop systems. Limmat [11] and Google Percolator [23] extend Hadoop by introducing a push-based processing, where data can be pushed into the process and results are computed computed incrementally on top of the current process state, e.g. aggregates for current windows. The main downside of these approaches is latency, which can run into minutes.

The Spark-Streaming [29] and Hadoop Online Prototype (HOP) [7] projects are an attempt to improve the Hadoop process by making it leaner and as a result, faster. When possible, data manipulation is performed directly in main memory without using secondary storage, which makes computation faster. Although these approaches improve performance for real-time analytics support, the micro-batch design creates an intrinsic limit that prevents these types of systems from achieving the same low latency as event-based systems.

Also included in the class of micro-batch systems, although not Map-Reduce oriented, is the Trident [28] system, an extension to Storm that provides higher level operators and other features. Trident suffers from the latency limitation mentioned previously for the micro-batch systems, a problem which we do not have in SDAP. Although both Spark-Streaming and Trident offer a set of predefined operators, developing a topology still requires the development of a program in Java or Scala, which unlike SDAP is a more challenging task because: users have to know the language; and users must validate their code before validating the application itself. SDAP enable designers, not developers, to rapidly develop topologies neglecting all details related to software code development and thus, focusing on the business logic. In addition, SDAP supports complex window definitions which are not available in any of these systems.

There has been much research on developing the performance of individual operators, e.g. [4,5,9,16–18,20,26]. In [12,13], the authors tackle the problem of processing XML data streams. They developed a multidimensional metamodel for constructing XML cubes to perform both direct recursion and indirect recursion analytics. While this approach has similar goals and approach, the SDAP system is designed to scale, adopts an easier to use scripting approach, and can facilitate JSON sources unlike their approach which only uses XML. In fact, none of these research efforts offer a comprehensive solution to maximize performance across all aspects of the streaming network. SDAP, on the other hand, provides a general solution for rapid development of stream analytics for high performance environments.

3   Streaming Analytics Case Study

Figure 1 illustrates a Bike Sharing stream analytics process design using the scenario described in Sect. 1. The data is streamed from bike sharing systems (BSS) in real-time from the cities of New York and Dublin. Here, the rounded rectangles represent the data manipulation steps. Arrows between steps describe how the stream flows from one transformation to the next. The operation applied by each step is depicted with a symbol (see the legend) within the rectangle, along with its degree of parallelism (within parenthesis). The specific operation performed by the operator is detailed with bold text just below each step. On top of each step, an italic text provides a brief explanation of the operation applied in the node. Note that the Selection operator outputs two streams: the solid edge denotes the stream of data satisfying the condition; the dashed edges

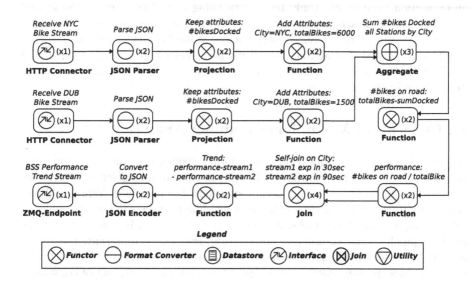

Fig. 1. Topology for bike sharing system case study in SDAP.

are the stream of data *not* satisfying the condition. Where one output stream from Selection is not used, the edge is not shown in the illustration.

The application in Fig. 1 describes an analysis of the bikes stream by generating the performance trend of the input BSS systems. Data flows into the application by the `HTTP Connector` step that connects to the BSS stream and delivers a snapshot of the status of all bike stations in each BSS system. Data is collected on a per minute basis. Station status data is provided in JSON format and thus, it is passed to the `JSON Parser` operator that convert data from JSON into the mapped tuple format. The next step removes attributes unnecessary for the required analysis at hand. Finally, two constants are added to the stream: the city the BSS data is from, and the number of bikes the BSS system has available. Streams from different cities are merged. As part of this process, there are three calculations.

- Available: the number of bikes currently docked across the city, as an aggregation of all docked bikes in each station in the incoming data over the interval of one minute (the data refresh rate);
- InUse: the number of bikes on the road, as the difference between the total bikes available and those docked;
- System Performance: defined as the ratio between the number of bikes on road and the total number of bikes (the more bikes on the road, the better the performance for this topology).

The remaining steps calculate the *trend* of the performance of each system. Performance trend is defined as the difference between two consecutive performances. In order to have two performances values in the same observation, a self-join (same city) is made on the performance stream, where one stream has an expiration time of 30 s (only fresh data being considered), while the second stream has an expiration time of 90 s. This way, a new performance value for a city is coupled with the previous performance value from the same city.

Once the trend is available, the result is converted into JSON format and produced in the output of the application, available to other applications, via a ZeroMQ end-point.

4   Conceptual Model for Streaming Analytics

In general, streaming applications consist of a sequence of data manipulation operations, where each operation performs a basic transformation to a data element, passing the result to the next operator in the sequence. When multiple transformations are chained together in a pipeline fashion, they create sophisticated, complex transformations. A *complex transformation* is a workflow, where *multiple* pipelines are combined. These workflows can be represented as direct acyclic graphs (DAGs) and are also referred to as topologies [4]. The SDAP model enables the construction of complex topologies using the set of constructs described below.

Tuple. A tuple is used to model any data element in a stream. It is composed of a list of values describing the occurrence of an event. For instance, in the BSS stream described later in Sect. 3, each update reports on the status of each station, where each station has an *identifier, address, status* (operative or not), the *number* of bikes docked, and *geo-location*.

Stream. A stream is a sequence of events described by tuples. Tuples in a stream conform to a (known) schema: each tuple value in the same stream are instances of a known set of attributes, each having a specific data type. For instance, tuples generated from bike stations status update on bike sharing system all have the same structure, with potentially different values, as people take and park bikes during the day.

Operator. An operator is a data processing step that processes each tuple received from one (or more) input stream(s) by applying a transformation to the tuple's data to generate a new tuple in the output stream. The operators are described in the following section but for now, we discuss two important parameters that are associated with each operator: *parallelism* and *protocol*.

> **Parallelism.** In a topology, each operator decides its degree of parallelism. Parallelism controls the number of instances of an operator that collaborate to complete a process. In order to process large amounts of data, processing must be distributed across multiple computational resources (cores, CPUs, machines).

> **Protocol.** The protocol defines how tuples are passed between the instances of contiguous operators in a topology. For example, a tuple can be passed to just one instance or to all instances of the next operator in the topology. SDAP supports four routing modes for protocol: *round-robin, direct, hash* and *broadcast*. In **round-robin** mode, tuples from an upstream node's output port are distributed to all instances, in an even fashion across all the downstream resources. **Direct** mode defines a direct and exclusive connection between *one* instance of the upstream node and *one* instance of the downstream node. This routing strategy is effective when pipelined operators require the same degree of parallelism. The **hash** mode routes tuples on the basis of a (key) value within the tuple itself. This permits an application to collect data having the same key in the same resource. Where this leads to uneven usage of downstream resources, the **broadcast** routing strategy, ensures that every tuple from a single instance of an upstream node, is copied to *all* instances of the downstream node.

Topology. A topology describes a stream analytics workflow, i.e. how the data stream flows from the input source(s) through the combination of primitive operators and sub-topologies to the output. It is modeled as a DAG, where nodes represent operators, and edges describes how tuples move between operators.

4.1 Primitive Operators in SDAP

This section presents a sample of the more important operators in SDAP, which are powerful enough to enable designers to construct very complex transformations. The rationale for providing a set of built-in operators is: (i) application designers focus on the transformation workflow and not implementation details; (ii) semantics are guaranteed and consistent across the entire system; (iii) every operator delivers the best possible performance; and (iv) the system can be extended with new operators as required. SDAP currently offers the following operators: Functor, Aggregate, Join, Sort, Interface, Format Converter, Datastore, Control and Utility.

The **Functor** operator applies a transformation that is confined and local to the tuple currently being processed. Many transformations can be thought as specializations of the Functor operator. SDAP provides *Projection* and *Selection*; *Function* which provides adding constants or a sequence attribute to the stream; text-to/from-date conversion; math (addition, division, modulo, etc.) and string functions.

The **Aggregate** operator groups tuples from the input stream, with an implementation of SQL-like aggregations: average, sum, max, min and count. The operator requires a *window* definition that specifies when and for how long tuples be included in the aggregation.

The **Join** operator is similar to the relational join but requires the definition of a window specifying the tuples from each stream to include in the join evaluation.

The **Sort** operator sorts the tuples within a "chunk" of the input stream in lexicographical order on the specified set of attributes. The number of tuples that comprise the chunk, is specified in a window definition.

The **Datastore** operator enables the stream to interact with a repository to retrieve, lookup, store and update data. The repository can be a database, a text file or an in-memory cache.

The **Interface** operator enables SDAP to create streams of data from external data sources to generate into topologies and to create end-points where processed data can be accessed by consumer applications. Consumer applications can be external or within SDAP (e.g. other topologies). Currently, SDAP can process streams from Twitter, Salesforce, ZeroMQ and generic HTTP endpoints.

The **FormatConverter** provides data format conversion between the tuple and other formats when processing data within a topology.

SDAP is extensible and new operators can be added as necessary. Currently, SDAP also includes the following additional operators: **Look-up**, to look-up values from either databases or files; **Geotagging**, to convert name of locations into geo-coordinates; **Delay**, to hold or slow down the elaboration of each tuple by some interval of time; **Heartbeat**, to signal all operators in a topology; **Cache**, to provide a fast memory space where to temporarily hold and share data across the whole topology; and **Tokenizer**, to transforms a text into multiple word tokens.

Some operators, e.g. join, cannot operate on an infinite stream of data: they require the definition of a *window* that cut the stream in "chunks" of data. Windows [4,5,18] are usually defined by specifying a set of constraints on attributes such as time, number of observed events (i.e. received tuples), or values in the stream(s) [16,20]. SDAP supports all of the above and, in addition, allows to define windows on sophisticated constraints involving conditions on both the input and output streams. Section 4.3 presents additional information and an example of a window definition.

4.2 Topology Model

A stream analytics process, called Topology, is modeled as a DAG, a directed acyclic graph, which is defined as in Definition 1:

Definition 1 (Topology). *A topology* $T = \langle N, R, \Sigma_O, \Sigma_R, \Sigma_P, o, r, p \rangle$ *is a five element tuple where* $N = \{n_1, n_2, \ldots\}$ *is a set of Nodes, and* $R = \{r_1, r_2, \ldots\}$ *is a set of Routes,* Σ_O *is the set of operators,* Σ_R *the set of data distribution protocols,* Σ_P *the degree of parallelism, and* o, r *and* p *functions that associates:* $o : N \rightarrow \Sigma_O$ *a node with an operator (and configuration),* $r : R \rightarrow \Sigma_R$ *an edge with a protocol, and* $p : N \rightarrow \Sigma_P$ *a node with a degree of parallelism.*

Each node in the topology specification follows the expression syntax outlined in Definition 2 in [3].

Definition 2 (Node)

```
operator
    <node-label>
    <operator-executable-path>
    <node-configuration-path>
```

Where: operator declares a node in the DAG; node-label specifies the label for such node, in order to refer to it in other places in the topology definition; operator-executable associates the executable with the node; finally, node-configuration specifies the arguments to pass to the executable and that configure the behaviour of the operator, e.g. conditions for a filtering criteria.

With reference to Fig. 1, Listing 1.1 shows an excerpt of the topology specification to illustrate how nodes in a DAG are declared in SDAP. The excerpt focuses on the top right part of Fig. 1, specifically on nodes: Keep Attributes: #bikesDocked, Add Attributes: CityNYC, total Bikes6000 and Sum #bikes Docked all Station by City. For convenience, above nodes are renamed to keep_attributes, add_constant, sum_docked, respectively.

Details on how to specify an operator's configuration are presented in the Sect. 4.3. In Listing 1.1, line 11 defines a node with label sum_docked, that is associated with operator Aggregation, whose configuration is in file

sum_docked_conf. It implements an aggregation operation, calculating a sum
of all docked bikes available at each station, grouping data by city.

```
1  ## Product stream, nodes
   ...
3  operator
       keep_attributes_dub
5      functor-mpi
       ${keep_attributes_dub_conf}
7  operator
       add_constants_dub
9      functor-mpi
       ${add_constants_dub_conf}
11 operator
       add_constants_nyc
13     functor-mpi
       ${add_constants_nyc_conf}
15 operator
       sum_docked
17     aggregate-mpi
       ${sum_docked_conf}
19 ...
```

Listing 1.1. Specification of nodes in a topology: an example.

The flow of tuples from operator to operator in the topology is defined along
with a routing protocol. Its specification follows Definition 3 from [3].

Definition 3 (Route)

```
route
    <upstream-node-label:port>
    <protocol>
    <downstream-node-label:port>
```

Where: route declares an edge in the topology; upstream-node-label is the
label of a node acting as data provider (also called upstream node); analogously,
downstream-node-label is the label of the other node participating in the
connection, specifically the label of the node receiving data (also called down-
stream node); protocol specifies how to distribute tuples between the two
nodes (e.g. *direct, round-robin, hash* or *broadcast*); port specifies which port
each node will use to send/receive tuples.

The specification in Listing 1.2 shows the part of the topology in Fig. 1
that links the *add_attributes_dub* and the *sum_docked* steps (also in Listing 1.1).
Specifically, all tuples from the add_attributes_dub node are passed to node
sum_docked via port number 1. Note that *sum_docked* receives data from two
upstream nodes, namely streams, one for the data from Dublin, the other for

the data form NYC. All edge declarations use the protocol `roundrobin` to
exchange tuples between the instances of the involved nodes.

```
## Product stream, connections
...
route
    add_attributes_dub:1 roundrobin add_constants_dub:1
route
    add_constants_dub:1 roundrobin sum_docked:1
route
    add_constants_nyc:1 roundrobin sum_docked:1
...
```

Listing 1.2. Specification of routes between nodes in a topology: an example.

Lastly, we need to associate each node with a degree of parallelism. The
syntax is the following, from [3]:

Definition 4 (Parallelism)

```
parallelism <node-label> <degree>
```

Where: `parallelism` declares the parallelism for a node; `node-label` indicates the node in question; and `degree` specifies the degree of parallelism, that
is how many runtime process instances have to be instantiated for the node in
question.

Listing 1.3 illustrate the final excerpt of the sample topology specification.
From Listing 1.3, we can see that nodes `keep_attributes`, `add_constants`,
and `sum_docked` have parallelism 2, 2, 3, respectively. The rationale in choosing
a degree of parallelism is based on the amount of data to process and on the cost of
the operation. In this example, the first two operations are rather simple, whether
the aggregation is actually performing a calculation, thus a higher degree of parallelism. Note that values in Listing 1.3 are for illustration purpose. Real-world
deployments these values have, in general, much higher values.

```
## Product stream, distribution
...
parallelism keep_attributes 2
parallelism add_constants 2
parallelism sum_docked 3
...
```

Listing 1.3. Specification of node parallelism: an example.

```
"in": [
   {"name": "timestamp", "type": "double"}
  ,{"name": "docked", "type": "int"}
  ,{"name": "City", "type": "String"}
  ,{"name": "totalBikes", "type": "int"}
  ]
,"out": [
   {"name": "timestamp", "type": "double"}
  ,{"name": "City", "type": "String"}
  ,{"name": "totalBikes", "type": "int"}
  ,{"name": "sumDockedBikes", "type": "int"}
  ]
...
```

Listing 1.4. Schema of input and output streams of a node.

```
...
,"groupby": [
   {"attribute": "City", "attribute": "totalBikes"}
  ]
,"aggregate": [
   {"input_attribute_name": "dockedBikes"
  ,"operation": "sum"
  ,"output_attribute_name": "sumDockedBykes"}
  ]
...
```

Listing 1.5. Detail of the aggregation operator configuration.

4.3   Operator Configuration

In a topology, a node is associated with an executable implementing a specific operator, e.g. Selection. The details of the nature of the input and output streams, as well as how to filter incoming tuples is provided in the operator configuration specification file. This specification starts with detailing the schemas of the input and output streams, that is attribute names and types. Then, each operator has its own signature, thus a different set of configuration parameters. Since it is not possible to illustrate the configuration of all operators in SDAP, we focus on just one of them: the Aggregate. Listings 1.4 and 1.5 shows excerpt of configuration for the aggregate operator in our example in Fig. 1. Listing 1.4 shows the schema of tuples for the input and the output streams. Listing 1.5 shows the details of aggregation, in this case a *sum*. It can be seen that values from the input streams are grouped by City (and totalBikes); the aggregations are defined on attribute dockedBikes; the results are provided in output attribute sumDockedBikes. An attribute timestamp is also added to the output stream.

   In contrast to the infinite nature of the data stream, the aggregation operator is required to work on a finite set of data. Finite sets of data are defined by *windows.* SDAP supports arbitrarily complex windows, including those based on: wall-clock intervals, number of observed tuples, the value of a progressing attribute [19] in the stream (e.g. time), external events (e.g. control messages), and conditional, that is based on values in the stream.

SDAP allows for the following window types: *Interrupt, Attribute*, and *Tuple*. With *Interrupt*, the window is defined by an external message: basically, the operator finalize the calculations and release the results only when requested. This options suits operations that have to release data at regular interval of (wall-clock) times. Type *Tuple* models windows defined on the number of input observations, e.g. create windows of 50 consecutive observations each, with a new window starting every 20. Type *Attribute* models window based on a progressing attribute embedded in the stream. The progressing attribute has the characteristic of being monotone in value, that is, increase at a standard interval (e.g. time).

When a basic window is not sufficient, developers can define window borders by condition: the developer can express an arbitrary condition to define when to close or open a new window. This is useful, for instance, to define landmark [10] windows or, more generally, windows whose boundaries depend on values in the data stream. SDAP allows developers to define conditions on attributes from both the input and output streams. An example of when a conditional window may be needed is the following: provide aggregate results immediately when the aggregate value exceeds a specified threshold defined by a literal in a constraint or by another value embedded in the stream. SDAP does not wait for windows to close to evaluate results: new partial, temporary, results are evaluated as new data is received. Thus, temporary results are always current and can be forwarded as part of a subsequent output.

Listing 1.6 continues the presentation of the configuration for the aggregation operator by specifying its window. The listing defines a tumbling window [10] on the progressing attribute *timestamp*. In fact we can see that: the window is of type `Attribute`; that the attribute characterizing the window is the `timestamp`; that a window should close (`window_close`) every 60 s; that values should be forwarded to output (`window_emit`) at the same time the window is closed (i.e. 60 s); that a new window should be created (`window_advance`) 60 s ahead of the previous open one; and that tuples are considered part of the "current" window if they arrive up to half a second after the specified window close limit.

```
 ...
2 ,"window_type":"Attribute"
 ,"progressing_attribute":"timestamp"
4 ,"window_close": {"type":"literal","size": 60}
 ,"window_emit": {"type":"literal","size": 60}
6 ,"window_advance": 60
 ,"window_delay": 0.5
8 ...
```

Listing 1.6. Specification of a tumbling window configuration.

5 SDAP System Architecture

While SDAP runs on a wide range of computational resources, the architecture was designed and implemented as a high performance system. In Fig. 2, the SDAP architecture is illustrated as having seven major components: the Resource Manager, Clustering, Data Operators, Monitoring, System Interface, Application Specification (repository) and the Resource Configuration repositories. Components such as the Resource Manager, Computation, Clustering and Monitoring use Slurm [25], MVAPICH2 [21], and Ganglia [8], as they are established, high performance open source source libraries.

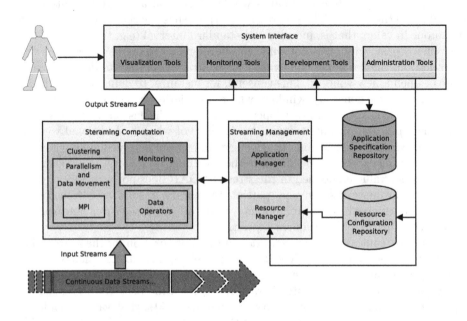

Fig. 2. SDAP architecture: logical view.

5.1 Data Operations

The Clustering component resides at the core of the SDAP architecture and comprises two sub-components: (i) parallelism and data movement and (ii) data operations. The first component manages parallel processes and the movement of data between processes. We adopt an implementation of the Message Passing Interface (MPI), specifically MVAPICH2, in order to optimize these high-performance environments. MPI is designed to achieve high performance, scalability, and portability and MVAPICH2 is one of the best performing implementation. This is mainly due to its support for the most recent and performing hardware, such as Infiniband [14], a high-performance inter-connector, designed to be scalable and featuring high throughput and low latency. The Parallelism and Data Movement component builds on top of Phish [24], that in turns uses MPI. Phish is library to develop topologies composed of executables, providing an abstraction on parallelism and message delivery. The Data Operations component implements the data processing operation and enforces the operation protocols across the parallel processes. Again using Phish, SDAP has a Data Operators component which offers a set of built-in operations, including selection, projection, join, etc.

5.2 Distribution Management

Figure 3 illustrate the physical architecture of SDAP. The cluster is divided in *Compute* and *Control* nodes. Compute nodes provide computation and are mutually independent while the control node manages and coordinates compute nodes. Specifically, each compute node hosts the operators' executables, to perform data manipulation, and is responsible for forwarding data to the next operator in the topology. The operators' executable are deployed to every compute node so that each node can accommodate any operation specified in a topology. Each compute node also hosts the slave processes of the resource manager and of the resource monitoring. The Resource Manager process maintains the state of available resources and the plan for allocation to each topology (e.g. a single core allocated exclusively or not to an operator of a topology). When the resource manager slave receives a request to allocate or to release a resource, it first checks the state of the resource and then applies requests where possible. The Resource Monitoring process collects resource usage data, i.e. CPU time, memory allocation, etc., for the local node. This data is then forwarded to the master node, where data is aggregated and evaluated. The master node dispatches resource allocation requests and analyzes the resource usage of all compute nodes. Topologies are deployed or recalled using the Application Manager component that, in turns, uses the *master* process of the resource manager to allocate the nodes as per topology specification, when possible. The resource manager Master Process collects and analyzes resource usage data sent by all slave processes residing on compute nodes. Resource usage is provided at both the individual and collective level: a user can analyze details of CPU, memory, and network load for each individual node or for the cluster as a whole. Resource load can be analyzed for any specified time interval.

5.3 Resource Management

As multiple streaming topologies run on a Cluster and since resources are limited, there is a requirement for managing and monitoring resources. The Components Resource Manager and Resource Monitoring components deliver on these requirements. The Resource Manager is built on top of Slurm [25], a high-performance, scalable and fault-tolerant cluster resource manager. Slurm provides functionality to execute, manage and monitor distributed parallel applications and is used in many of the most powerful computers in the world. It

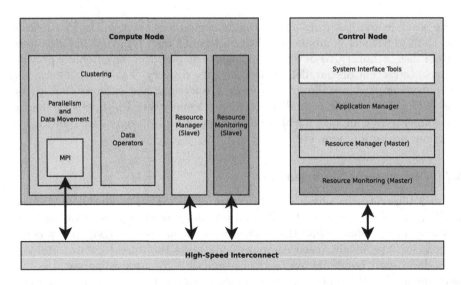

Fig. 3. SDAP architecture: physical view.

facilitates computational resource management and the allocation and deployment of streaming topologies. Among its features are topology relocation (to other resources) and a fault-tolerance mechanism. We have integrated Ganglia [8] as our resource monitoring system as it is highly scalable and works in high-performance environments.

Because the SDAP system was designed for the highest levels of scalability and performance, the resource monitoring and manager components can be deployed in a hierarchical manner, as illustrated in Fig. 4. Such a hierarchical organization of processes facilitates resource allocation requests and monitoring to be distributed over a larger number of processes and thus, avoiding bottlenecks at either the CPU or network level.

For data monitoring and analysis, the processes between the bottom and top of the hierarchy can perform partial aggregations which further reduces the load on the control node.

The System Interface component includes tools such as: the development environment, result visualization and monitoring and administrative tools. The Application Specification repository maintains all defined topologies, allowing users to store, retrieve and update topology specifications. Finally, the Resource Configuration maintains the configuration of the resources available on the computational cluster.

6   Experiments

In this section, we present our evaluation of SDAP, which consists of two parts: performance, and usability. Performance evaluation focuses on the ability of SDAP to deliver low latency data processing at scale. The usability evaluation

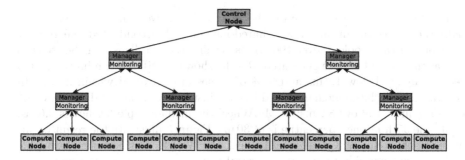

Fig. 4. SDAP architecture: resource and monitor manager scalability.

focuses on the simplicity of use of the tool, compared to popular alternative systems.

6.1 Performance

Latency is a crucial metric for streaming data in a high-performance environment and is defined as the interval of time between the solicitation and the response of a system. For systems targeting analytics on big data, it is important to maintain low latency when the inputs, and subsequent resources, grow to a large number.

We compare SDAP with Apache Storm [27] (version 1.0.2, latest version available at the time of writing) and Google's MillWheel [1]. These systems have been chosen because they adopt the event based data processing paradigm, as with SDAP. In particular, the former explicitly targets low latency performance at a scale; while the latter focused initially on the provision of event based processing primitives and scalability, and has now evolved to deliver low latency in its recent versions.

The common ground on which to compare the systems is a topology composed of the following steps: a data generator, followed by a non blocking stream operation (e.g. select), and a collector. The data generator step generates random data tuples of about 100B each. In implementing the non blocking operation, the stream operation step perform the following tasks: (i) record the timestamp of when the tuple is received, (ii) scan all attributes in the tuple, to emulate an operator worst case scenario where the operation needs to access all data, (iii) attach a timestamp to the tuple, and (iv) forward the tuple in output to the collector step. The collector records the timestamp of the tuple arrival. The two timestamp are used to calculate the intrinsic latency of the system. Specifically, it is the time elapsed between the reception of the input (system solicitation), and the execution and delivery of the data manipulation to the next step in the topology (system response).

Experiments were conducted on the CUNY's High Performance Computing center. Each node is equipped with 2.2 GHz Intel Sandybridge processors with 12 cores, has 48 GB of RAM and uses the Mellanox FDR interconnect. The topology is deployed so that contiguous steps in the topology require inter-host communication and thus, require the use of the network media to exchange data (i.e. no communication via shared-memory). We conducted the test by distributing the topology over 100 CPUs, scaling the parallelism of the steps as well as the amount of data generated. Results show that SDAP exhibits a median record latency below 1.2 ms and 95% of tuples are delivered within 2.1 ms. In contrast: Storm has a median record latency just below 3.2 ms, and delivers 95% of tuples in just above 3.2 ms; in MillWheel the median tuple latency is 3.6 ms, while the 95th latency percentile is of 30 ms. Figure 5 illustrate the result of our experiment, excluding Google's MillWheel. MillWheel's platform is not available so it was not possible to run an empirical test: our comparison and analysis is based on the author's evaluation in [1].

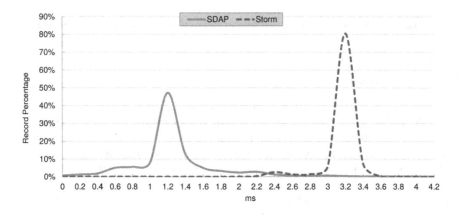

Fig. 5. Tuple processing and delivery latency.

It can be seen that SDAP processes data faster that Storm (and MillWheel) but the latency values are spread close to the median. With respect to Storm, SDAP performs about 3 times better for both the median and the 95th percentile latency. In comparison with MillWheel, SDAP performs 3 times better on the median latency and 10 times better on the 95th percentile latency. Overall, SDAP performs between 200% and 900% better than the other systems.

We also conducted a test to verify how the latency changes when the same number of execution processes are distributed over a small number of machines machines, compared to a large number. We have prepared 4 test scenarios, namely Set8, Set24, Set48, and Set96, see [3]. Each scenario is run on different machine numbers, from the lowest to the highest number of machines that can accommodate the test. For instance, scenario Set8 requires 8 processes with processes run as follows: on the same machine with 8 CPUs; on two machines, using 4 CPUs from each; on four machines, using 2 CPUs from each; and eight machines, using just one CPU from each. For Set96 we started with 12 machines using 8 CPU from each, down to 96 machines using 1 CPU only from each. This was repeated for the remaining configurations. The result of running these scenarios is illustrated in Fig. 6 from [3].

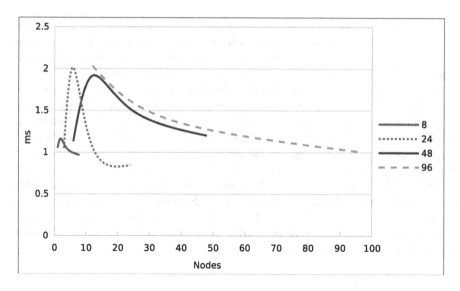

Fig. 6. Tuple processing and delivery latency time by node usage in SDAP.

It can be seen that: (i) the latency performance is quite stable across all configurations, with values supporting the results from the previous experiment; and (ii) the best configuration is when all processes are grouped together on the same machines or when they are highly distributed across different machines. The latter can be as follows: latency is low when all processes are grouped on the same (few) machine(s) because data transfer is (mostly) performed via shared memory (i.e. not via network); latency is also low when the least amount of CPUs is used per machine, because not enough data is exchanged via the network interface which as a result, does not become saturated. Latency is higher when an intermediate number of CPUs are used per machine because the processes generate enough data to flood the network interface while not being able to take advantage of data exchange via shared memory.

6.2 Ease of Development

In this section, we discuss SDAP ease of development, that is, the effort required to develop and maintain a topology. Since MillWheel is not available, and Storm does not provide built-in operators, we have decided to compare SDAP with another popular system: Spark-Streaming. Spark-Streaming provides built-in operators and allows the designer to specify stream applications in a quite succinct manner as Scala programs. While Spark-Streaming supports other languages, Scala has been chosen because it is one of the less verbose and is supported natively.

Let us compare the two systems using a streaming application that must detect tuples that match a set of specified keywords keywordSeq. If a tuple contains a target keyword, it is then forwarded in JSON format to a Kafka end-point. Listing 1.7 shows such application for Spark-Streaming. As we can see, specifying operations such as the cartesian product is rather straightforward. However, even simple operations require a rather verbose specification. To begin with, the developer must select the right libraries to use, such as what package to use for the JSON conversion – omitted in the snippet. Since there are multiple possibilities, the developer is required to study each alternative to determine which one best fits her needs, which takes time. Then, the developer must compose the application. Let us ignore the details of the streams, i.e. the attributes. After the cartesian product operation, line 3, and before checking the keyword match, line 15, the developer must manually open multiple connections to Kafka for each node of the Spark-Streaming cluster. Specifically, the developer opens a connection for each partition of data in the resilient distributed dataset (or RDD, the main data structure in Spark), in an attempt to parallelize the data exchange between the two systems.

```
...
2  statuses.foreachRDD(rdd => {  //  for each RDD
   val cartesian = rdd.cartesian(keywordSeq)
4
   //  for each partition of data, connect to the end-point
6  cartesian.foreachPartition(partitionOfRecords => {
   //  initialize the Kafka producer
8      val props = new HashMap[String, Object]()
       setupKafkaProps(props)
10     val producer = new KafkaProducer[String, String](props)
   //  for each record in partition, check keyword match
12   partitionOfRecords.foreach{
       case (status, keywords) => {
14 //     if a keyword matches, forward to end-point
       if (keywords.map(l => l.toLowerCase()).toSet
16        subsetOf status.toLowerCase().split(" ").toSet) {
         val jsonMessage = ("text" -> record.toString)
18         ~ ("keywords" -> keywords.toList)
         val jsonMessageString = compact(render(jsonMessage))
20 //        send message to kafka
         val message = new ProducerRecord[String, String]
22          (topicsOutputSet.head, null, jsonMessageString)
         producer.send(message)
24 ...
```

Listing 1.7. Keyword match sample application in spark-streaming.

In contrast, SDAP: (i) has no need to study libraries for inclusion as they are built-in; (ii) the cartesian operator can also be expressed simply but requires no knowledge of a programming language, Scala in this case; (iii) the connection to the end-point is provided by a built-in operator that does not require the developer to study the inner workings of Scala optimization for Spark-Streaming; and (iv) the set comparison between the record value and a set of keywords can be implemented as a sequence of tokenizer + selection operators. In total, the SDAP would have 4 operators and associated configuration files. Note that the configuration files would be mostly empty, and the in/out stream attributes are automatically populated using the designer portal. Listing 1.8 shows the equivalent topology specification with details of configuration files omitted for the sake of space.

```
...
operator cross_product join-mpi ${join_conf}
operator tokenize_keyword utility-mpi ${tokenizer_conf}
operator keyword_match functor-mpi ${selection_conf}
operator json_encoder_converter-mpi ${json_conf}
operator kafka_endpoint interface-mpi ${kafka_conf}
...
route cross_product:1 roundrobin tokenize_keyword_set:1
route tokenize_keyword_set:1 roundrobin keyword_match:1
route keyword_match:1 roundrobin json_endpoint:1
route keyword_json:1 roundrobin kafka_endpoint:1
```

Listing 1.8. Keyword match equivalent application in SDAP.

It can be observed that the SDAP implementation is easier to read and does not require any previous programming knowledge. In our experience with the SDAP, we have observed that users rapidly familiarize with topology paradigm, with the options of the operators and become power-users capable of developing rather complex transformations.

7   Conclusions

The increasing availability of data provided through online channels has led to an increasing demand to include this form of data in many decision making processes for growing numbers of organizations. The increasing volumes of this data means a greater need for high performance streaming processors. Current systems have been shown to suffer from issues of latency and/or overly complex design and implementation methods. SDAP provides the capability to design and deploy topologies which can scale to very high volumes of data while hiding the complexities of these systems from the designer. Its powerful operators provide a platform for highly complex analytics with SDAP abstracting the underlying management of data and parallel processing. Our evaluation shows SDAP to outperform popular streaming systems such as Storm and MillWheel. Our current research is focused on a few fronts: analysis of application patters, optimization of resource usage, and performance. On one side, we want to exploit the declarative nature of the approach to further simplify the design of stream analytics, and to discover application and resource optimization opportunity.

The visibility and ease of access to the data transformations operation allows to analyze stream analytics design patterns and to optimize the resource allocation. On the other side, we want to further improve performance of the execution engine by including hardware acceleration, e.g. using graphics processing units (GPUs), in the logic of the operators in the context of a high-performance and low-latency environment.

Acknowledgements. This research was supported, in part, from Collective[i] Grant RF-7M617-00-01, the National Science Foundation Grants CNS-0958379,CNS-0855217, ACI-1126113 and the City University of New York High Performance Computing Center at the College of Staten Island.

References

1. Akidau, T., Balikov, A., Bekiroglu, K., Chernyak, S., Haberman, J., Lax, R., McVeety, S., Mills, D., Nordstrom, P., Whittle, S.: Millwheel: fault-tolerant stream processing at internet scale. PVLDB **6**(11), 1033–1044 (2013). http://www.vldb.org/pvldb/vol6/p1033-akidau.pdf
2. Balazinska, M., Balakrishnan, H., Madden, S., Stonebraker, M.: Fault-tolerance in the borealis distributed stream processing system. ACM Trans. Database Syst. **33**(1), 1–3 (2008). http://doi.acm.org/10.1145/1331904.1331907
3. Cappellari, P., Chun, S.A., Roantree, M.: Ise: a high performance system for processing data streams. In: Proceedings of 5th International Conference on Data Science, Technology and Applications, DATA 2016, Lisbon, Portugal, pp. 13–24, 24–26 July 2016
4. Carney, D., Çetintemel, U., Cherniack, M., Convey, C., Lee, S., Seidman, G., Stonebraker, M., Tatbul, N., Zdonik, S.B.: Monitoring streams - a new class of data management applications. In: Proceedings of 28th International Conference on Very Large Data Bases, VLDB 2002, Hong Kong, China, pp. 215–226, 20–23 August 2002. http://www.vldb.org/conf/2002/S07P02.pdf
5. Chandrasekaran, S., Franklin, M.J.: Streaming queries over streaming data. In: Proceedings of 28th International Conference on Very Large Data Bases, VLDB 2002, Hong Kong, China, pp. 203–214, 20–23 August 2002. http://www.vldb.org/conf/2002/S07P01.pdf
6. Chen, X., Beschastnikh, I., Zhuang, L., Yang, F., Qian, Z., Zhou, L., Shen, G., Shen, J.: Sonora: a platform for continuous mobile-cloud computing. Technical report (2012). https://www.microsoft.com/en-us/research/publication/sonora-a-platform-for-continuous-mobile-cloud-computing/
7. Condie, T., Conway, N., Alvaro, P., Hellerstein, J.M., Gerth, J., Talbot, J., Elmeleegy, K., Sears, R.: Online aggregation and continuous query support in mapreduce. In: Proceedings of the ACM SIGMOD International Conference on Management of Data, SIGMOD 2010, Indianapolis, Indiana, USA, pp. 1115–1118, 6–10 June 2010. http://doi.acm.org/10.1145/1807167.1807295
8. Ganglia (2015). http://ganglia.sourceforge.net/. Accessed 15 Nov 2016
9. Gedik, B., Yu, P.S., Bordawekar, R.: Executing stream joins on the cell processor. In: Proceedings of the 33rd International Conference on Very Large Data Bases, University of Vienna, Austria, pp. 363–374, 23–27 September 2007. http://www.vldb.org/conf/2007/papers/research/p363-gedik.pdf

10. Gehrke, J., Korn, F., Srivastava, D.: On computing correlated aggregates over continual data streams. In: Mehrotra, S., Sellis, T.K. (eds.) Proceedings of the 2001 ACM SIGMOD International Conference on Management of Data, Santa Barbara, CA, USA, pp. 13–24. ACM, 21–24 May 2001. http://doi.acm.org/10.1145/375663.375665
11. Grinev, M., Grineva, M.P., Hentschel, M., Kossmann, D.: Analytics for the real-time web. PVLDB **4**(12), 1391–1394 (2011). http://www.vldb.org/pvldb/vol4/p1391-grinev.pdf
12. Gui, H., Roantree, M.: Topological XML data cube construction. Int. J. Web Eng. Technol. **8**(4), 347–368 (2013)
13. Gui, H., Roantree, M.: Using a pipeline approach to build data cube for large XML data streams. In: Hong, B., Meng, X., Chen, L., Winiwarter, W., Song, W. (eds.) DASFAA 2013. LNCS, vol. 7827, pp. 59–73. Springer, Heidelberg (2013). doi:10.1007/978-3-642-40270-8_5
14. Infiniband (2015). http://www.infinibandta.org/. Accessed 15 Nov 2016
15. InfoSphere streams (2015). http://www-03.ibm.com/software/products/en/infosphere-streams. Accessed 15 Nov 2016
16. Kang, J., Naughton, J.F., Viglas, S.: Evaluating window joins over unbounded streams. In: Proceedings of the 19th International Conference on Data Engineering, Bangalore, India, pp. 341–352, 5–8 March 2003. doi:10.1109/ICDE.2003.1260804
17. Li, J., Maier, D., Tufte, K., Papadimos, V., Tucker, P.A.: Semantics and evaluation techniques for window aggregates in data streams. In: Proceedings of the ACM SIGMOD International Conference on Management of Data, Baltimore, Maryland, USA, pp. 311–322, 14–16 June 2005. http://doi.acm.org/10.1145/1066157.1066193
18. Madden, S., Shah, M.A., Hellerstein, J.M., Raman, V.: Continuously adaptive continuous queries over streams. In: Proceedings of the 2002 ACM SIGMOD International Conference on Management of Data, Madison, Wisconsin, pp. 49–60, 3–6 June 2002. http://doi.acm.org/10.1145/564691.564698
19. Maier, D., Li, J., Tucker, P., Tufte, K., Papadimos, V.: Semantics of data streams and operators. In: Eiter, T., Libkin, L. (eds.) ICDT 2005. LNCS, vol. 3363, pp. 37–52. Springer, Heidelberg (2004). doi:10.1007/978-3-540-30570-5_3
20. Motwani, R., Widom, J., Arasu, A., Babcock, B., Babu, S., Datar, M., Manku, G.S., Olston, C., Rosenstein, J., Varma, R.: Query processing, approximation, and resource management in a data stream management system. In: CIDR (2003). http://www-db.cs.wisc.edu/cidr/cidr2003/program/p22.pdf
21. MVAPICH2, The Ohio State University (2015). http://mvapich.cse.ohio-state.edu/. Accessed 15 Nov 2016
22. Neumeyer, L., Robbins, B., Nair, A., Kesari, A.: S4: distributed stream computing platform. In: Proceedings of the 2010 IEEE International Conference on Data Mining Workshops, ICDMW 2010, Washington, DC, USA, pp. 170–177 (2010). IEEE Computer Society. doi:10.1109/ICDMW.2010.172
23. Peng, D., Dabek, F.: Large-scale incremental processing using distributed transactions and notifications. In: Proceedings of the 9th USENIX Symposium on Operating Systems Design and Implementation, OSDI 2010, Vancouver, BC, Canada, pp. 251–264, 4–6 October 2010. http://www.usenix.org/events/osdi10/tech/full_papers/Peng.pdf
24. Plimpton, S.J., Shead, T.M.: Streaming data analytics via message passing with application to graph algorithms. J. Parallel Distrib. Comput. **74**(8), 2687–2698 (2014). doi:10.1016/j.jpdc.2014.04.001
25. Slurm (2015). http://slurm.schedmd.com/. Accessed 15 Nov 2016

26. Teubner, J., Müller, R.: How soccer players would do stream joins. In: Proceedings of the ACM SIGMOD International Conference on Management of Data, SIGMOD 2011, Athens, Greece, pp. 625–636, 12–16 June 2011. http://doi.acm.org/10.1145/1989323.1989389

27. Toshniwal, A., Taneja, S., Shukla, A., Ramasamy, K., Patel, J.M., Kulkarni, S., Jackson, J., Gade, K., Fu, M., Donham, J., Bhagat, N., Mittal, S., Ryaboy, D.V.: Storm@twitter. In: International Conference on Management of Data, SIGMOD 2014, Snowbird, UT, USA, pp. 147–156, 22–27 June 2014. http://doi.acm.org/10.1145/2588555.2595641

28. Trident (2012). http://storm.apache.org/documentation/Trident-tutorial.html. Accessed 15 Nov 2016

29. Zaharia, M., Das, T., Li, H., Shenker, S., Stoica, I.: Discretized streams: an efficient and fault-tolerant model for stream processing on large clusters. In: 4th USENIX Workshop on Hot Topics in Cloud Computing, HotCloud 2012, Boston, MA, USA, 12–13 June 2012. https://www.usenix.org/conference/hotcloud12/workshop-program/presentation/zaharia

Identifying Conversational Message Threads by Integrating Classification and Data Clustering

Giacomo Domeniconi[1]([✉]), Konstantinos Semertzidis[2], Gianluca Moro[1],
Vanessa Lopez[3], Spyros Kotoulas[3], and Elizabeth M. Daly[3]

[1] Department of Computer Science and Engineering (DISI),
University of Bologna at Cesena, Cesena, Italy
{giacomo.domeniconi,gianluca.moro}@unibo.it
[2] Department of Computer Science and Engineering, University of Ioannina,
Ioannina, Greece
ksemer@cs.uoi.gr
[3] IBM Research - Damastown Industrial Estate Mulhuddart, Dublin 15, Ireland
{vanlopez,Spyros.Kotoulas,elizabeth.daly}@ie.ibm.com

Abstract. Conversational message thread identification regards a wide
spectrum of applications, ranging from social network marketing to virus
propagation, digital forensics, etc. Many different approaches have been
proposed in literature for the identification of conversational threads
focusing on features that are strongly dependent on the dataset. In this
paper, we introduce a novel method to identify threads from any type of
conversational texts overcoming the limitation of previously determining
specific features for each dataset. Given a pool of messages, our method
extracts and maps in a three dimensional representation the semantic
content, the social interactions and the timestamp; then it clusters each
message into conversational threads. We extend our previous work by
introducing a deep learning approach and by performing new extensive
experiments and comparisons with classical learning algorithms.

Keywords: Conversational message · Thread identification · Data clustering · Classification

1 Introduction

Nowadays, online conversations have become widespread, such as email, web
chats, online conversations and social groups. Online chatting, is a fast, economical and efficient way of sharing information and it also provides users the ability
to discuss different topics with different people. Understanding the context of
digital conversations finds a wide spectrum of applications such as marketing,
social network extraction, expert finding, the improvement of email management,
ranking content and others [1–4].

G. Domeniconi—This work was partially supported by the european project "TOREADOR" (grant agreement no. H2020-688797).

C. Francalanci and M. Helfert (Eds.): DATA 2016, CCIS 737, pp. 25–46, 2017.
DOI: 10.1007/978-3-319-62911-7_2

The contiguous increase of digital content leads people being overwhelmed by information. For example, imagine the case where a user has hundreds of new unread messages in a chat or a mailbox or in a situation where the same user needs to track and organise posts in forums or social groups. In order to instantly have a clear view of different discussions, avoiding expensive and tedious human efforts, we need to automatically organise this data stream into threads.

Many different approaches have been proposed in the related literature to extract topics from document sets, mainly through a variety of techniques derived from Probabilistic Latent Semantic Indexing (pLSI) [5] and Latent Dirichlet Allocation (LDA) [6]. However, the problem of identifying threads from conversational messages differs from document topic extraction for several aspects [4,7–10]: (i) conversational messages are generally much shorter than usual documents making the task of topic identification much more difficult (ii) thread identification strongly depends on social interactions between the users involved in a message exchange, (iii) as well the time of the discussion.

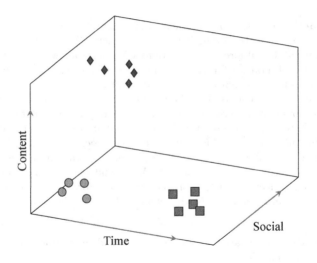

Fig. 1. Three dimensional representation of threads messages.

In our previous work [11] we addressed the problem of efficiently identifying conversational threads from pools of online messages - for example from emails, social groups, chats etc. In other words, we looked for the sets of messages that are related to each other with respect to text content, time and involved users.

We consider a three dimensional representation [12] which consists of text content, temporal information, and social relations. In Fig. 1, we depict the three dimensional representation which illustrates 3 threads with different colours and shapes, that yields to total of 14 messages. The green circles and red squares threads have the same social and content dimensions but not time. While the blue diamonds thread consists of different topics and users, but it occurs in the same time frame of the green circles one. The use of the three dimensional representation leads to emphasis of thread separation.

We propose several measures to exploit the messages features, based on this three dimensional representation. Then, similarly to the work in [13], the generated features are embedded into a metric distance in density and hierarchical clustering algorithms [14,15] which cluster messages in threads. In order to enhance our approach to efficiently identify threads in any type of dataset, we train a classification model with a set of messages previously organised in threads. The classifiers exploit the same features used in the clustering phase and they return the probability that a pair of messages belong to the same thread. In other words, a binary supervised model is trained with instances, each referring to a pair of messages. Each instance uses the same features described previously, and a label describing whether the two messages belong to the same thread or not. This model provides a probability of being in the same thread for a pair of messages, we propose to use this probability as a similarity distance in clustering methods to identify the threads. We observe that the classifiers' output can help the clustering process to achieve higher accuracy by identifying the threads correctly. In this paper we extend our aforementioned approach by comparing classical Machine Learning supervised algorithms with a Deep Learning Multi-Layer Perceptron. Deep learning algorithms are proven to achieve good results in several mining domains [16], especially in big-data contexts [17]; thus considering that the identification of conversational thread in social networks could be a problem with an huge amount of data to analyze, a deep learning approach could fit well into this task.

We have extensively evaluated our approach with real world datasets including emails and social group chats. Our experimental results show that our method can identify the large majority of the threads in several type of dataset, such as web conversation including emails, chats and posts.

To summarize, the main contributions of this work are:

- a three dimensional message representation based on textual semantic content, social interactions and time to generate features for each message;
- clustering algorithms to identify threads, on top of the features generated from the three dimensional representation;
- combination of the generated features to build classifiers that identify the membership probability of pair of messages to the same thread and this probability is used as a distance function for the clustering methods to identify threads;
- extension of the combined classification techniques with clustering algorithms that achieves a higher accuracy than using clustering alone;
- comparison of performances obtained by several machine learning algorithms and the deep learning Multi-Layer Perceptron.

The rest of this paper is structured as follows. In Sect. 4, we present related work, while in Sect. 3, we formally define the thread identification problem. In Sect. 2, we introduce our model and our algorithms for thread identification.

Section 3.3 presents the experimental results on real datasets and Sect. 5 concludes the paper.

2 Method

In this section, we outline a generic algorithm for identifying messages which belong to the same thread from a set of messages $\mathcal{M}$, such as emails, social group posts and chats. As an intermediate step, the algorithm addresses the problem of computing the similarity measure between pairs of messages. We propose a suite of features and two methods to combine them (one unsupervised and one supervised) to compute the similarity measure between two messages. We also present clustering algorithms which identify threads based on this similarity measure in Sect. 2.3.

2.1 Data Model

We consider a set of messages $\mathcal{M} = \{m_1, m_2, ...\}$ that refers to online texts such as emails, social group chats or forums. Each message is characterized by the following properties: (1) textual data (content and subject in case of emails), (2) creation time, and (3) the users involved (authors or sender/recipients in case of emails). We represent each message as a three-dimensional model [12,18] to capture all these components. Thus, a message $m \in \mathcal{M}$ can be denoted as a triplet $m = <c_m, \mathcal{U}_m, t_m>$, where c_m refers to text content, $\mathcal{U}_m = \{u_1, u_2, ...\}$ refers to the set of users that are involved in m, and t_m refers to the creation time. Some dimensions can be missing, for instance chat, groups and forum messages provide only the author information, without any recipients.

A conversation thread is defined as a set of messages exchanged on the same topic among the same group of users during a time interval, more formally, the set of messages $\mathcal{M}$ is partitioned in a set of conversations $\mathcal{C}$. Each message $m \in \mathcal{M}$ belongs to one and only one conversation $c \in \mathcal{C}$. The goal of the thread reconstruction task is to automatically identify the conversations within a pool of messages. To this aim, we propose a clustering-based method that relies on a similarity measure between a pair of messages, called $SIM(m_i, m_j)$. In the following sections, we define different proposed approaches to calculate the similarity measure. In the rest of the paper, we will use the notation $\Omega = \{\omega_1, \omega_2, ...\}$ to refer the predicted extracted conversations.

2.2 Messages Features

Social text messages, like emails or posts, can be summarized by three main components: text content, temporal information, and social relations [12]. Each of the three main components can be analyzed under different points of view to compute the distance between a pair of messages, which involves the creation of several features. The function $SIM(m_i, m_j)$ relies on these features and returns a similarity value for each pair of messages (m_i, m_j), which is used by the clustering algorithm that returns the finding threads. We now present the extracted features used to measure the similarity between two messages.

The content component relies on the semantics of the messages. There are two main sources: the messages text and the subject, if present (e.g., social network posts do not have this information). The first considered feature is the similarity of the messages text content. We make use of the common *Bag of Words* (*BoW*) representation, that describes a textual message m by means of a vector $\mathcal{W}(m) = \{w_1, w_2, ...\}$, where each entry indicates the presence or absence of a word w_i. Single words occurring in the message text are extracted, discarding punctuation. A stopwords list is used to filter-out all the words that are not informative enough. The standard Porter stemming algorithm [19] is used to group words with a common stems. To estimate the importance to each word, there exist several different weighting schemes [20], here we make use of the commonly used *tf.idf* scheme [21].

Using *BoW* representation, the similarity between two vectors m_i, m_j can be measured by means of the commonly used *cosine similarity* [22]:

$$f_{C_T}(m_i, m_j) = \frac{\mathcal{W}(m_i) \cdot \mathcal{W}(m_j)}{\|\mathcal{W}(m_i)\| \|\mathcal{W}(m_j)\|}$$

Since by definition the *BoW* vectors have only positive values, the $f_{C_T}(m_i, m_j)$ takes values between zero and one, being zero if the two vectors do not share any word, and one if the two vectors are identical. In scenarios where the subject is available, the same process is carried out, computing the similarity cosine $f_{C_S}(m_i, m_j)$ of words contained in the messages subject.

The cosine similarity allows a lexical comparison between two messages but does not consider the semantic similarity between two messages. There are two main shortcomings of this measure: the lack of focus on keywords, or semantic concepts expressed by messages, and the lack of recognition of lexicographically different words but with similar meaning (i.e. synonyms), although this is partially computed through the stemming. In order to also handle this aspect, we extend the text similarity by measuring the correlation between entities, keywords and concepts extracted using AlchemyAPI[1]. AlchemyAPI is a web service that analyzes the unstructured content, exposing the semantic richness in the data. Among the various information retrieved by AlchemyAPI, we take into consideration the extracted topic keywords, involved entities (e.g. people, companies, organizations, cities and other types of entities) and concepts which are the abstractions of the text (for example, "My favorite brands are BMW and Porsche = Automotive industry"). These three information are extracted by Alchemy API with a confidence value ranging from 0 to 1. We create three vectors, one for each component of the Alchemy API results for keywords, entities and concepts for each message and using the related confidence extracted by AlchemyAPI as weight. Again we compute the cosine similarity of these vectors, creating three novel features:

---

[1] http://www.alchemyapi.com/.

- $f_{C_K}(m_i, m_j)$: computes the cosine similarity of the keywords of m_i and m_j. This enables us to quantify the similarity of the message content based purely on keywords rather than the message as a whole.
- $f_{C_E}(m_i, m_j)$: computes the cosine similarity of the entities that appear in m_i and m_j focusing on the entities shared by the two messages.
- $f_{C_C}(m_i, m_j)$: computes the cosine similarity of the concepts in m_i and m_j, allowing the comparison of the two messages on a higher level of abstraction: from words to the expressed concepts.

The second component is related to the social similarity. For each message m, we create a vector of involved users $\mathcal{U}(m) = \{u_1, u_2, ...\}$ defined as the union of the sender and the recipients of m (note that the recipients information is generally not provided in social network posts). We exploit the social relatedness of two messages through two different features:

- The similarity of the users involved in the two messages $f_{S_U}(m_i, m_j)$, defined as the Jaccard similarity between $\mathcal{U}(m_i)$ and $\mathcal{U}(m_j)$:

$$f_{S_U}(m_i, m_j) = \frac{|\mathcal{U}(m_i) \cap \mathcal{U}(m_j)|}{|\mathcal{U}(m_i) \cup \mathcal{U}(m_j)|}$$

- The neighborhood Jaccard similarity $f_{S_N}(m_i, m_j)$ of the involved users. The neighborhood set $\mathcal{N}(u)$ of an user u is defined as the set of users that have received at least one message from u. We also include each user u in its neighborhood $\mathcal{N}(u)$ set. The neighborhood similarity of two messages m_i and m_j is defined as follows:

$$f_{S_N}(m_i, m_j) = \frac{1}{|\mathcal{U}(m_i)||\mathcal{U}(m_j)|} \sum_{\substack{u_i \in \mathcal{U}(m_i) \\ u_j \in \mathcal{U}(m_j)}} \frac{|\mathcal{N}(u_i) \cap \mathcal{N}(u_j)|}{|\mathcal{N}(u_i) \cup \mathcal{N}(u_j)|}$$

Finally, the last component relies on the time of two messages. We define the time similarity as the logarithm of the inverse of the distance between the two messages, expressed in days, as follows:

$$f_T(m_i, m_j) = \log_2(1 + \frac{1}{1 + |t_{m_i} - t_{m_j}|})$$

We use the inverse normalization of the distance in order to give a value between zero and one, where zero correspond to a high temporal distance and one refers to messages with low distance.

As a practical example, Fig. 2 shows two messages, with the related properties, and the values of the features generated from them.

2.3   Clustering

In this section, we present the clustering methods used to identify the threads. Based on the set of aforementioned features $\mathcal{F} = \{f_{C_T}, f_{C_S}, f_{C_K},$

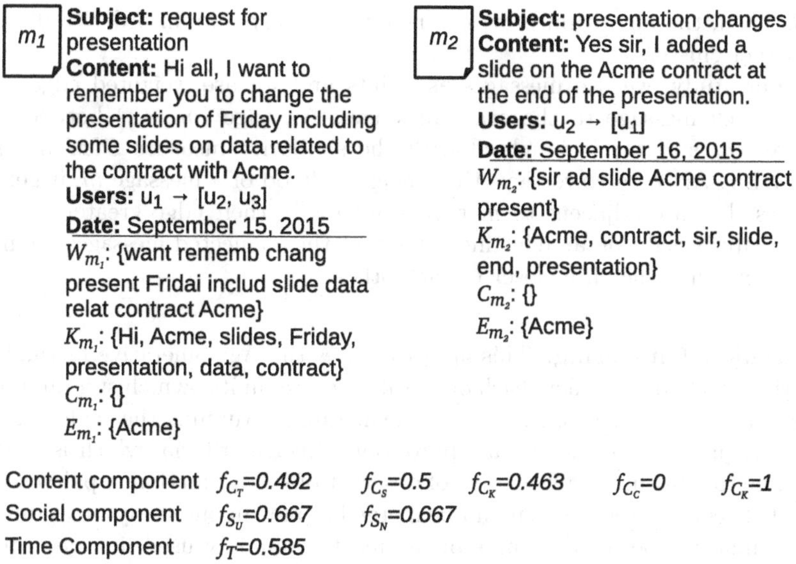

m_1 **Subject:** request for presentation **Content:** Hi all, I want to remember you to change the presentation of Friday including some slides on data related to the contract with Acme.
Users: $u_1 \rightarrow [u_2, u_3]$
Date: September 15, 2015
W_{m_1}: {want rememb chang present Fridai includ slide data relat contract Acme}
K_{m_1}: {Hi, Acme, slides, Friday, presentation, data, contract}
C_{m_1}: {}
E_{m_1}: {Acme}

m_2 **Subject:** presentation changes **Content:** Yes sir, I added a slide on the Acme contract at the end of the presentation.
Users: $u_2 \rightarrow [u_1]$
Date: September 16, 2015
W_{m_2}: {sir ad slide Acme contract present}
K_{m_2}: {Acme, contract, sir, slide, end, presentation}
C_{m_2}: {}
E_{m_2}: {Acme}

Content component f_{C_T}=0.492 f_{C_S}=0.5 f_{C_K}=0.463 f_{C_C}=0 f_{C_K}=1
Social component f_{S_U}=0.667 f_{S_N}=0.667
Time Component f_T=0.585

Fig. 2. Example of features calculation for a pair of messages. Message components: Subject, Content, Users (sender → recipients) and creation date. $\mathcal{W}(m_i)$ refers to the bag of words of a message obtained after the tokenization, stopwords removal and stemming. The vectors of keywords ($\mathcal{K}(m_i)$), concepts ($\mathcal{C}(m_i)$) and entities ($\mathcal{E}(m_i)$) extracted from AlchemyAPI are shown. In the bottom the values for each proposed feature are also shown. For simplicity, we assume binary weight for components.

$f_{C_E}, f_{C_C}, f_{S_U}, f_{S_N}, f_T\}$, we define a distance measure that quantifies the similarity between two messages:

$$SIM(m_i, m_j) = \Pi_{f \in \mathcal{F}}(1 + f(m_i, m_j)) \qquad (1)$$

We compute a $N \times N$ matrix with the similarities between each pair of messages (m_i, m_j) and we use density based and hierarchical clustering algorithms, being the two most common distance-based approaches.

Density-Based Clustering. We use the DBSCAN [14] density-based clustering algorithm in order to cluster messages to threads because given a set of points in some space, DBSCAN groups points that are closely packed together (with many nearby neighbors). DBSCAN requires two run time parameters, the minimum number min of points per cluster, and a threshold θ that defines the neighborhood distance between points in a cluster. The algorithm starts by selecting an arbitrary point, which has not been visited, and by retrieving its θ-neighborhood it creates a cluster if the number of points in that neighborhood is equals to or greater than min. In situations where the point resides in a dense part of an existing cluster, its θ-neighbor points are retrieved and are added to the cluster. This process stops when the densely-connected cluster is completely found. Then, the algorithm processes new unvisited points in order to discover any further clusters.

In our study, we use messages as points and we use weighted edges that connect each message to the other messages. An edge (m_i, m_j) between two messages m_i and m_j is weighted with the similarity measure $SIM(m_i, m_j)$. When DBSCAN tries to retrieve the θ-neighborhood of a message m, it gets all messages that are adjacent to m with a weight in their edge greater or equal to θ. Greater weight on an edge indicates that the connected messages are more similar, and thus they are closer to each other.

Hierarchical Clustering. This approach uses the Agglomerative hierarchical clustering method [15] where each observation starts in its own cluster, and pairs of clusters are merged as one moves up the hierarchy. Running the agglomerative method requires the choice of an appropriate linkage criteria, which is used to determine the distance between sets of observations as a function of pairwise distances between clusters that should be merged or not. In our study we examined, in preliminary experiments, three of the most commonly used linkage criteria, namely the *single, complete* and *average linkage* [23]. We observed that average linkage clustering leads to the best results. The average linkage clustering of two clusters of messages Ω_y and Ω_z is defined as follows:

$$avgLinkCl(\Omega_y, \Omega_z) = \frac{1}{|\Omega_y||\Omega_z|} \sum_{\substack{\omega_i \in \Omega_y \\ \omega_j \in \Omega_z}} SIM(\omega_i, \omega_j)$$

The agglomerative clustering method is an iterative process that merges the two clusters with highest average linkage score. After each merge of the clusters, the algorithm starts by recomputing the new average linkage scores between all clusters. This process runs until a cluster pair exists with a similarity greater than a given threshold.

2.4   Classification

The clustering algorithms described above rely on the similarity measure SIM, that combines with a simple multiplication several features, to obtain a single final score. This similarity measure in Eq. 1 gives the same weight, namely importance, to each feature. This avoids the requirement to tune the parameters related to each feature, but could provide an excessively rough evaluation and thus bad performance. A different possible approach, is to combine the sub components of similarity measure SIM as features into a binary supervised model, in which each instance refers to a pair of messages, the features are the same described in the Sect. 2.2 and the label is one if the messages belonging to the same thread and zero otherwise. At runtime, this classifier is used to predict the probability that two messages belong to the same thread, using this probability as the distance between the pairs of messages into the same clustering

Table 1. Characteristics of datasets.

Dataset	Messages type	#messages	#threads	#users	Peculiarities
BC3	Emails	261	40	159	Threads contain emails with different subject
Apache	Emails from mailing list	2945	334	113	Threads always contain emails with same subject
Redhat	Emails from mailing list	12981	802	931	Threads always contain emails with same subject
WhoWorld	Posts from Facebook page	2464	132	1853	Subject and recipients not available
Healthy Choice	Posts from Facebook page	1115	132	601	Subject and recipients not available
Healthcare Advice	Posts from Facebook group	3436	468	801	Subject and recipients not available
Ireland S. Android	Posts from Facebook group	4831	408	354	Subject and recipients not available

algorithms. The benefit of such approach is that it automatically finds the appropriate features to use for each dataset and it leads to a more complete view of the importance of each feature. Although it is shown in [24] that decision trees are faster and more accurate in classifying text data, we experimented with a variety of classifiers.

The classification requires a labeled dataset to train a supervised model. The proposed classifier relies on data in which each instance represents a pair of messages. Given a set of training messages $\mathcal{M}_{Tr}$ with known conversation subdivision, we create the training set coupling each training message $m \in \mathcal{M}_{Tr}$ with n_s messages of $\mathcal{M}_{Tr}$ that belong to the same thread of m and n_d messages belonging to different threads. We label each training instance with one if the corresponding pair of messages belong to same thread and zero otherwise. Each of these coupled messages are picked randomly. Theoretically we could create $(|\mathcal{M}_{Tr}| \cdot |\mathcal{M}_{Tr}| - 1|)/2$ instances, coupling each message with the whole training set. In preliminary tests using *Random Forest* as the classification model, we notice that coupling each training message with a few dozen same and different messages can attain higher performances. All the experiments are conducted using $n_s = n_d = 20$, i.e. each message is coupled with at maximum 20 messages of the same conversation and 20 of different ones. In the rest of the paper we refer to the proposed clustering algorithm based on a supervised model, as *SVC*.

As it will be shown in the Sect. 3.3, the Agglomerative hierarchical clustering achieves better results with respect to the DBSCAN, thus, we use this clustering algorithm in the *SVC* approach.

2.5 Multi-layer Perceptron

A Multi-Layer Perceptron (MLP) is a deep learning algorithm [16,25] that can be viewed as a logistic regression classifier where the input is first transformed using a learnt non-linear transformation Φ. This transformation maps the input data into a space where they expect to be better linearly separable. This intermediate layer is referred to as a hidden layer.

Formally, a one-hidden-layer MLP is a function $f : R^D \to R^L$, where D is the size of input vector x and L is the size of the output vector $f(x)$, such that, in matrix notation:

$$f(x) = G(b^{(2)} + W^{(2)}(s(b^{(1)} + W^{(1)}x))),$$

with x that is the input vector, i.e. the set of feature values $\mathcal{F}$ computed for the coupled messages; $b^{(1)}$, $b^{(2)}$ are bias vectors; $W^{(1)}$, $W^{(2)}$ are weight matrices and G and s activation functions. The vector $h(x) = \Phi(x) = s(b^{(1)} + W^{(1)}x)$ constitutes the hidden layer. $W^{(1)} \in R^{D \times D_h}$ is the weight matrix connecting the input vector to the hidden layer. Each column $W^{(1)}_{\cdot i}$ represents the weights from the input units to the i-th hidden unit. After a pre-tuning phase, we chose the *hyperbolic tangent* activation function:

$$s = tanh(a) = (e^a - e^{-a})/(e^a + e^{-a})$$

for its faster and higher results with respects to other functions, like for instance the *sigmoid*. The output vector is then obtained as:

$$o(x) = G(b^{(2)} + W^{(2)}h(x))$$

By considering that we need a binary classification, namely either the coupled messages belong or not to the same thread, the output is a couple of probabilities for each class achieved by choosing G as the *softmax* function. To run our experiments, we make use of Yusuke java implementation Sugomori[2] [26].

3 Evaluation

In this section, we compare the accuracy of the clustering methods described in Sect. 2 in terms of identifying the actual threads.

3.1 Datasets

For evaluating our approach we consider the following seven real datasets:

– The *BC3* dataset [27], which is a special preparation of a portion W3C corpus [28] that consists of 40 conversation threads. Each thread has been annotated by three different annotators, such as extractive summaries, abstractive summaries with linked sentences, and sentences labeled with speech acts, meta sentences and subjectivity.

---

[2] https://github.com/yusugomori/DeepLearning.

- The *Apache* dataset which is a subset of Apache Tomcat public mailing list[3] and it contains the discussions from August 2011 to March 2012.
- The *Redhat* dataset which is a subset of Fedora Redhat Project public mailing list[4] and it contains the discussions that took place in the first six months of 2009.
- Two Facebook pages datasets, namely *Healthy Choice*[5] and *World Health Organizations*[6], crawled using the Facebook API[7]. They consist of real posts and relative replies between June and August 2015. We considered only the text content of the posts (discarding links, pictures, videos, etc.) and only those written in English (AlchemyAPI is used to identify the language).
- Two Facebook public groups datasets, namely *Healthcare Advice*[8] and *Ireland Support Android*[9], also crawled using the Facebook API. They consist of conversations between June and August 2015. Also for this dataset we considered only the text content of the posts written in english.

We use the first three datasets that consist of emails in order to compare our approach with existing related work [29–31] on conversation thread reconstruction in email messages. To our knowledge, there are no publicly available datasets of social network posts with a gold standard of conversation subdivision. We use the four Facebook datasets to evaluate our method in a real social network domain.

The considered datasets have different peculiarities, in order to evaluate our proposed method under several perspectives. *BC3* is a quite small dataset (only 40 threads) of emails, but with the peculiarity of being manually curated. In this dataset is possible to have emails with different subjects in the same conversation. However, in *Apache* and *Redhat* the messages in the same thread, have also the same subject.

With regards to Facebook datasets, we decided to use both pages and groups. Facebook pages are completely open for all users to read and comment in a conversation. In contrast, only the members of a group are able to view and comment a group post and this leads to a peculiarity of different social interaction nets. Furthermore, each message - post - in these datasets has available only the text content, the sender and the time, without information related to subject and recipients. Thus, we do not take into account the similarities that use the recipients or subject. Table 1 provides a summary of the characteristics of each dataset.

In the experiments requiring a labeled set to train a supervised model, the datasets are evaluated with 5-fold cross-validation, subdividing each of those in 5 thread folds.

---

[3] http://tomcat.apache.org/mail/dev.
[4] http://www.redhat.com/archives/fedora-devel-list.
[5] https://www.facebook.com/healthychoice.
[6] https://www.facebook.com/WHO.
[7] https://developers.facebook.com/docs/graph-api.
[8] https://www.facebook.com/groups/533592236741787.
[9] https://www.facebook.com/groups/848992498510493.

3.2 Evaluation Metrics

The *precision*, *recall* and F_1-*measure* [23] are used to evaluate the effectiveness of the conversation threads identification. Here, we explain these metrics in the context of the conversational identification problem. We evaluate each pair of messages in the test set. A true positive (TP) decision correctly assigns two similar messages to the same conversation. Similarly, a true negative (TN) assigns two dissimilar messages to different threads. A false positive (FP) case would be when the two messages do not belong to the same thread but are labelled as co-threads in the extracted conversations. Finally, false negative (FN) case is when the two messages belong to the same thread but are not co-threads in the extracted conversations. Precision (p) and recall (r) are defined as follows:

$$p = \frac{TP}{TP + FP} \qquad\qquad r = \frac{TP}{TP + FN}$$

The F_1-*measure* is defined by combining the precision and recall together, as follows:

$$F_1 = \frac{2 \cdot p \cdot r}{p + r}$$

We also use the *purity* metric to evaluate the clustering. The dominant conversation, i.e. the conversation with the highest number of messages inside a cluster, is selected from each extracted thread cluster. Then, purity is measured by counting the number of correctly assigned messages considering the dominant conversation as cluster label and finally dividing by the number of total messages. We formally define purity as

$$purity(\Omega, \mathcal{C}) = \frac{1}{|\mathcal{M}|} \sum_k \max_j |\omega_k \in c_j|$$

where $\Omega = \{\omega_1, \omega_2, ..., \omega_k\}$ is the set of extracted conversations and $\mathcal{C} = \{c_1, c_2, ..., c_j\}$ is the set of real conversations.

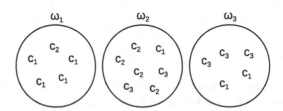

Fig. 3. Conversation extraction example. Each ω_k refers to an extracted thread and each c_j corresponds to the real conversation of the message.

To better understand the purity metric, we refer to the example of thread identification depicted in Fig. 3. For each cluster, the dominant conversation and the number of related messages are: $\omega_1 : c_1, 4$, $\omega_2 : c_2, 4$, $\omega_3 : c_3, 3$. The total number of messages is $|\mathcal{M}| = 17$. Thus, the purity value is calculated as $purity = (4 + 4 + 3)/17 = 0.647$.

A final measure of the effectiveness of the clustering method, is the simple comparison between the the the number of identified threads ($|\Omega|$) against the number of real conversations ($|\mathcal{C}|$).

3.3   Results

Tables 2 and 3 report the results obtained with the seven datasets using the Weka [32] implementation of *Random Forest* algorithm. We applied a 2×2 cost matrix with a weight of 100 for the instances labelled with one and the *Sugomori* Java implementation of the *Multi-Layer Perceptron* as described above. The reported results are related to the best tuning of the threshold parameter of the clustering approaches, both for DBSCAN and Agglomerative. Further analysis on the parameters of our method are discussed in the next section.

Table 2 shows the results on the email datasets, on which we can compare our results (SVC) with other existing approaches, such as the studies of Wu and Oard [31], Erera and Carmel [30] and the lastest one of Dehghani et al. [29]. The first two approaches [30,31] are unsupervised, as the two clustering baselines, while the approach in [29] is supervised, like our proposed SVC; both this supervised methods are evaluated with the same 5-fold cross-validation, described above. All of the existing approaches use the information related to the subject of the emails, we show in the top part of the table a comparison using also the subject as feature in our proposed approach. We want point out that in *Apache* and *Redhat* dataset, the use of the subject could make the clusterization effortless, since all messages of a thread have same subject. It is notable how our supervised approach obtains really high results, reaching almost perfect predictions and always outperforming the existing approaches, particularly in *Redhat* and *Apache* dataset.

In our view, the middle of Table 2 is of particular interest, where we do not considered the subject information. The results, especially in *Redhat* and *Apache*, have a little drop, remaining anyhow at high levels, higher than all existing approaches that take into consideration the subject. Including the subject or not, the use of a supervised model to evaluate the similarity between two messages, brings a great improvement to the clustering performances, compared to the use of a simple combination of each feature as described in Sect. 2.3. In the middle part of Table 2 is also shown the effectiveness of our SVC predictor without the three features related to AlchemyAPI information; these features lead to an improvement of results especially in *Redhat*, which is the largest and more challenging dataset.

Table 2 also compares the performances of the SVCmethod using both the Random Forest algorithm (SVC_{RF}) and the Multi-Layer Perceptron (SVC_{MLP}). Noteworthy is the drop of precision and recall of the MLP algorithm for the apache and redhat dataset when the subject are not considered, in this case the Random Forest clearly outperforms the Deep Learning approach.

Table 2. Conversational identification results on email datasets. [30,31], DBSCAN and Agglom. are unsupervised methods, while [29] and SVC are supervised, both using the *Random Forest* (RF) and the *Multi-Layer Perceptron* (MLP) algorithms. The top part of the table shows the results obtained by methods using subject information, the middle part shows those achieved without such feature, finally the bottom part shows the results obtained with SVC method considering only a single dimension. With + and − we indicate respectively the use or not of the specified feature (s: subject feature, a: the three Alchemy features). For clustering and SVC approach we report results with best threshold tuning.

Methods	BC3			Apache			Redhat		
	Precision	*Recall*	F_1	*Precision*	*Recall*	F_1	*Precision*	*Recall*	F_1
Wu and Oard [31]	0.601	0.625	0.613	0.406	0.459	0.430	0.498	0.526	0.512
Erera and Carmel [30]	0.891	0.903	0.897	0.771	0.705	0.736	0.808	0.832	0.82
Dehghani et al. [29]	0.992	0.972	0.982	0.854	0.824	0.839	0.880	0.890	0.885
DBSCAN (+s)	0.871	0.737	0.798	0.359	0.555	0.436	0.666	0.302	0.416
Agglom. (+s)	**1.000**	0.954	0.976	0.358	0.918	0.515	0.792	0.873	0.83
SVC_{RF} (+s)	**1.000**	**0.986**	**0.993**	0.998	**1.000**	**0.999**	0.995	**0.984**	**0.989**
SVC_{MLP} (+s)	**1.000**	0.982	0.991	0.969	**1.000**	0.984	**0.99**	0.981	0.985
DBSCAN (−s)	0.696	0.615	0.653	0.569	0.312	0.403	0.072	0.098	0.083
Agglom. (−s)	**1.000**	0.954	0.976	0.548	0.355	0.431	0.374	0.427	0.399
SVC_{RF} (−s)	**1.000**	0.952	0.975	**0.916**	0.972	**0.943**	0.966	**0.914**	**0.939**
SVC_{RF} (−s −a)	0.967	**0.979**	0.973	0.892	**0.994**	0.940	0.815	0.699	0.753
SVC_{MLP} (−s)	**1.000**	0.973	**0.986**	0.654	0.671	0.663	0.703	0.541	0.612
SVC_{MLP} (−s −a)	**1.000**	0.972	0.986	0.633	0.616	0.624	0.469	0.579	0.518
SVC_{RF} (content)	**1.000**	**0.919**	**0.958**	0.954	**0.974**	**0.964**	0.988	**0.984**	**0.986**
SVC_{RF} (content −s)	0.964	0.902	0.932	0.604	0.706	0.651	0.899	0.872	0.885
SVC_{RF} (content −s −a)	**1.000**	0.828	0.905	0.539	0.565	0.552	0.68	0.558	0.613
SVC_{RF} (social)	0.939	0.717	0.813	0.345	0.361	0.353	0.360	0.045	0.08
SVC_{RF} (time)	0.971	0.897	0.933	0.656	0.938	0.772	0.376	0.795	0.511

Table 3. Conversation identification results on Facebook post datasets (subject and recipient information are not available). The top part of the table shows the results obtained considering all the dimensions, both using the *Random Forest* (RF) and the *Multi-Layer Perceptron* (MLP) algorithms; the bottom part shows the results obtained with SVC method considering only a single dimension. For clustering and our approach we report results with best threshold tuning.

Methods	Healthy Choice			World Health Org.			Healthcare Advice			Ireland S. Android		
	Precision	*Recall*	F_1	*Precision*	*Recall*	F_1	*Precision*	*Recall*	F_1	*Precision*	*Recall*	F_1
DBSCAN	0.027	0.058	0.037	0.159	0.043	0.067	0.206	0.051	0.082	0.201	0.002	0.004
Agglom	0.228	0.351	0.276	0.154	0.399	0.223	0.429	0.498	0.461	0.143	0.141	0.142
SVC_{RF}	**0.670**	0.712	**0.690**	0.552	0.714	0.623	**0.809**	0.721	0.763	0.685	0.655	0.67
SVC_{RF} (−a)	0.656	0.713	0.683	0.543	**0.742**	0.627	0.802	0.733	**0.766**	**0.708**	0.714	**0.711**
SVC_{MLP}	0.657	0.722	0.688	**0.615**	0.698	**0.654**	0.665	0.762	0.71	0.68	**0.739**	0.709
SVC_{MLP} (−a)	0.65	**0.726**	0.686	0.61	0.696	0.65	0.664	**0.764**	0.71	0.68	0.737	0.708
SVC_{RF} (content)	0.308	0.032	0.058	0.406	0.120	0.185	0.443	0.148	0.222	0.127	0.042	0.063
SVC_{RF} (content −a)	0.286	0.025	0.046	0.376	0.11	0.171	0.414	0.127	0.195	0.105	0.033	0.050
SVC_{RF} (social)	0	0	0	0	0	0	0.548	0.188	0.280	0.155	0.234	0.186
SVC_{RF} (time)	**0.689**	**0.670**	**0.679**	0.531	**0.750**	**0.622**	0.638	**0.769**	0.697	0.667	0.703	0.685

The aforementioned considerations, are valid also for the experiments on social network posts. To the best of our knowledge, there is not any related work on such type of datasets. In Table 3, we report the results of our approach on the four Facebook datasets. These data do not provide the subject and recipients information of messages, thus the reported results are obtained without the features related to the subject and neighborhood similarities, namely $f_{C_S}(m_i, m_j)$ and $f_{S_N}(m_i, m_j)$. We notice that the pure unsupervised clustering methods, particularly DBSCAN, achieve low *precision* and *recall*. This is due to the real difficulties of these post's data: single posts are generally short with little semantic information. For example suppose we have two simultaneous conversations *t1*: *"How is the battery of your new phone?"* - *"good!"* and *t2*: *"how was the movie yesterday?"* - *"awesome!"*. By using only the semantic information of the content, it is not possible to associate the replies to the right question, thus the time and the social components become crucial. Although there is a large amount of literature to handle grammatical errors or misspelling, in our study we have not taken into account these issues. Despite these difficulties, our method guided by a supervised model achieves quite good results in such data, with an improvement almost always greater than 100% with respect the pure unsupervised clustering. Results in Table 3 show the difficulties also for AlchemyAPI to extract valuable information from short text posts. In fact, results using the AlchemyAPI related features does not lead to better results.

The results achieved by the SVC method for each dimension are reported at the bottom of the Tables 2 and 3, in particular those regarding the content dimension have been produced with all features, excepts the subject and the Alchemy related features. In Table 2 is notable that considering the content dimension together with the subject feature leads, as expected, to the highest accuracy. By excluding the subject feature, SVC produces quite good results with each dimension, however they are lower than those obtained by the complete method; this shows that the three dimensional representation leads to better clusterisation.

Table 3 shows the differentiation in the results related to the Facebook datasets. In particular, the social dimension performs poorly if used alone, in fact the author of a message is known, whereas not the receiver user; also the text content dimension behaves badly if considered alone. In these datasets, the time appears to be the most important feature to discriminate the conversations, however the results achieved only with this dimension are worse than those of the SVC complete method.

From these results, achieved using each dimension separately from the others, we deduce that SVC is robust to different types of data. Moreover the use of a supervised algorithm allows both to identify the importance of the three dimensions and to achieve a method that can deal with different datasets without requiring ad-hoc tuning or interventions.

Parameter Tuning. Parameter tuning in machine learning techniques is often a bottleneck and a crucial task in order to obtain good results. In addition, for practical applications, it is essential that methods are not overly sensitive to parameter values. Our proposed method requires the setting of few parameters. In this section, we show the effect of changing different parameter settings. A first investigation of our SVC regards the supervised algorithm used to define the similarity score between a pair of messages.

We conducted a series of experiments on the benchmark datasets varying the model. Namely, we used decision trees (*Random Forest*), SVM (*LibSVM*), *Logistic Regression* and Deep Learning *Multi-Layer Perceptron*. For all the standard Machine Learning algorithms we used the default parameter values provided by the Weka implementation. For the MLP we done a tuning - we omit for space reasons - and we defined and fixed in all the experiments the following parameters: (i) *hyperbolic tangent* activation function: $s = tanh(a)$, (ii) number of training iterations: $n_e = 500$, (iii) number of hidden layers: $n_h = 50$, and (iv) learning rate: $lr = 0.5$.

Considering the intrinsic lack of balance of the problem (i.e. each message has a plenty of pairs with messages that belong to different threads and just few in the same one) we also experimented with a cost-sensitive version of Random Forest, setting a ratio of 100 for instances with messages belonging to the same thread. Table 4 shows the results, it is notable that the cost sensitive Random Forest always outperforms the standard Random Forest. Logistic regression and

Table 4. Results varying the supervised model used to compute the distance between two email.

Model	Purity	Precision	Recall	F_1	$|\Omega|$
BC3 ($|\mathcal{C}| = 40$)					
LibSVM	0.980	0.962	0.984	0.973	40
Logistic	**1.000**	**1.000**	0.965	**0.982**	45
RF	**1.000**	**1.000**	0.961	0.980	45
RF:100	**1.000**	**1.000**	0.952	0.975	46
MLP	**1.000**	**1.000**	0.973	0.986	44
Apache ($|\mathcal{C}| = 334$)					
LibSVM	0.785	0.584	0.583	0.584	500
Logistic	0.883	0.904	0.883	0.893	275
RF	0.862	0.885	**0.979**	0.930	255
RF:100	**0.920**	**0.916**	0.972	**0.943**	286
MLP	0.821	0.654	0.671	0.663	431
Redhat ($|\mathcal{C}| = 802$)					
LibSVM	0.575	0.473	0.674	0.556	450
Logistic	0.709	0.619	0.697	0.656	572
RF	0.89	0.888	0.900	0.894	762
RF:100	**0.954**	**0.966**	**0.914**	**0.939**	818
MLP	0.773	0.703	0.541	0.612	820
Facebook page: **Healty Choice** $|\mathcal{C}| = 132$)					
LibSVM	0.766	0.657	0.694	0.675	187
Logistic	0.788	0.676	**0.724**	**0.699**	211
RF	0.771	**0.682**	0.656	0.668	218
RF:100	0.787	0.670	0.712	0.690	214
MLP	**0.792**	0.657	0.722	0.688	220
Facebook page: **World Health Organization** ($n_c = 132$)					
LibSVM	0.628	0.444	**0.805**	0.573	118
Logistic	0.755	0.566	0.702	0.627	198
RF	0.731	0.536	0.718	0.614	186
RF:100	0.747	0.552	0.714	0.623	222
MLP	**0.784**	**0.615**	0.698	**0.654**	220
Facebook group: **Healthcare Advice** ($|\mathcal{C}| = 468$)					
LibSVM	0.692	0.502	0.768	0.607	383
Logistic	0.840	0.699	0.761	0.729	548
RF	0.766	0.596	**0.773**	0.673	467
RF:100	**0.909**	**0.809**	0.721	**0.763**	714
MLP	0.822	0.665	0.762	0.71	531
Facebook page: **Ireland Support Android** ($|\mathcal{C}| = 408$)					
LibSVM	0.655	0.460	0.744	0.568	356
Logistic	0.814	0.654	0.723	0.687	573
RF	0.786	0.646	0.641	0.644	627
RF:100	0.821	**0.685**	0.655	0.670	663
MLP	**0.837**	0.68	**0.739**	**0.709**	583

cost sensitive Random Forest achieve better results, with a little predominance of the latter.

An interesting outcome deduced by analysing Table 4 is that, as already reported in [33], the Random Forest algorithm obtains really good results, particularly when one is able to dive it into the right space of features opportunely weighted with respect to the classes. The deep learning method should overcome this problem, in fact MLP obtains higher performance in 6 out 7 datasets with respect to the simple unweighted Random Forest, but only in 1 out of 7 datasets are instead better of the cost-sensitive RF. For this reason, the choice of one or other algorithm highly depends on the previous knowledge available about the dataset.

The main parameter of our proposed method regards the threshold value used in the clustering algorithms. We experimented with the use of a supervised model in the DBSCAN clustering algorithm, but we noticed the results were not good. This is not surprising if we consider how DBSCAN works: it groups messages in a cluster iteratively adding the neighbors of the messages belonging to the cluster itself. This leads to the erroneous merge of two different conversations, if just one pair of messages is misclassified as similar, bringing a sharp decline to the clustering precision. The previous issue, however, does not affect the agglomerative clustering, because of the use of average link of two messages inside two clusters, to decide whether to merge them or not. In this approach the choice of the threshold parameter is crucial, namely the stop merge criterion. Figure 4 shows the F_1 trend varying the agglomerative threshold, using the weighted Random Forest as the supervised model. Is notable that all the trends have only one peak that corresponds to a global maximum, thus with a simple gradient descent is possible to find the best threshold value. Furthermore, our method is generally highly effective for threshold values ranging from 0.1 to 0.3, as shown

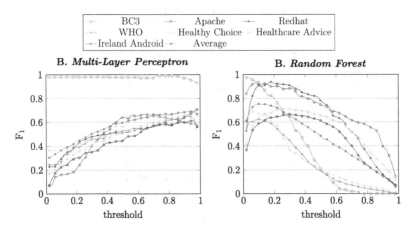

Fig. 4. F_1 measure for varying number of threshold, using (**A**) the MLP-based supervised algorithm and (**B**) the Random Forest algorithm.

in Fig. 4. This is also confirmed by the average trend, that has a peak with a threshold equal to 0.1.

4   Related Work

In last years, thread identification has received a lot of attention, including content and metadata based approaches. Metadata based approaches refers to header fields that are contained in emails or forum posts (e.g. send-to, reply-to). Content based approaches focus on text analysis on subject and content text. In this paper, we differentiate from the existing works by generalizing the problem of identifying threads in different types of datasets, not only in email sets like the most of related work [10,30,31,34,35]. The authors of [31] focus on identifying conversational threads from emails using only the subject. They cluster all messages with the same subject and at least one participant in common. Here, we also handle cases where messages belong to the same thread but have different subject. Similarly, in [34] the authors identify threads in emails using the extracted header information. They first try to identify the parent/child relationships using Zawinski algorithm[10] and then they use a topic-based heuristic to merge or decompose threads to conversations. Another approach for identifying threads in emails is proposed in [30], where clustering into threads exploits a similarity function that considers all relevant email attributes, such as subject, participants, text content and date of creation. Quotations are taken into account in [10] where combined with several heuristics such as subject, sender/recipient relationships among email and time, and as a result can construct email threads with high precision. Emails relationships are also considered in [35] where the authors use a segmentation and identification of duplicate emails and they group them together based on reply and forwarding relationships.

The work most closely related to ours is that of [29], that studies the conversation tree reconstruction, by first identifying the threads from a set of emails. Specifically, they map the thread identification problem to a graph clustering task. They create a semantic network of a set of emails where the nodes denote emails and the weighted edges represent co-thread relationships between emails. Then, they use a clustering method to extract the conversation threads. However, their approach is focus only on email datasets and their results are strongly bound with the used features, since when they do not take into account all features they have a high reduction in their accuracy. In contrast here, we consider general datasets and by using our classification model we are able to identify threads even when there are missing features. Although, it is not clear which graph clustering algorithm is used and how it identifies the clusters. We conduct an extensive comparison between our approach and the study of [29] in Sect. 5.

Another line of research addresses mining threads from online chats [4,7–9]. Specifically, the study of [4] focuses on identifying threads of conversation by using pattern recognition techniques in multi-topic and multi-person chat-rooms. In [9] they focus on conversation topic thread identification and extraction in a

---

[10] https://www.jwz.org/doc/threading.html.

chat session. They use an augmented *tf.idf* to compute weights between messages' texts as a distance metric exploiting the use of Princeton WordNet[11] ontology, since related messages may not include identical terms, they may in fact include terms that are in the same semantic category. In combination with the computed distance between messages they use the creation time in order to group messages with high similarity in a short time interval. In [7], they propose three variations of a single-pass clustering algorithm for exploiting the temporal information in the streams. They also use an algorithm based on linguistic features in order to exploit the discourse structure information. A single-pass clustering algorithm is also used in [8] which employs the contextual correlation between short text streams. Similar to [9], they use the concept of correlative degree, which describes the probability of the contextual correlation between two messages, and the concept of neighboring co-occurrence, which shows the number features co-existing in both messages.

Finally, there also exists a line of research on reconstructing the discussion tree structure of a thread conversation. In [36], a probabilistic model in conditional random fields framework is used to predict the replying structure for online forum discussions. The study in [24] employs conversation threads to improve forum retrieval. Specifically, they use a classification model based on decision trees and given a variety of features, including creation time, name of authors, quoted text content and thread length, which allows them to recover the reply structures in forum threads in an accurate and efficient way. The aforementioned works achieve really high performance (more than 90% of accuracy) in the conversation tree reconstruction, while the state of the art in threads identification obtains lower performance, about 80% for emails data and 60% for chats and short messages data. To this end, in this study we focus on improving thread identification performance.

5   Conclusions

This paper has studied the problem of identifying threads from a pool of messages that may correspond to social network chats, mailing list, email boxes, chats, forums etc. We have addressed the problem by introducing a novel method which given a pool of messages, it leverages the textual semantic content, the social interactions and the creation time in order to group the messages into threads. The work contains an analysis of features extracted from messages and it presents a similarity measure between messages, which is used in clustering algorithms that map messages to threads. Moreover the paper introduces a supervised model that combines the extracted features together with the probability of couples of messages to belong to the same thread, which is interpreted as a distance measure between two messages. Experiments show that this method leads to higher accuracy in thread identification, outperforming all earlier approaches.

Furthermore we investigated two differents main supervised approaches to create the classification model: standard machine learning algorithms, such as

---

[11] http://wordnet.princeton.edu/.

Random Forest, SVM and Logistic regression, and the deep learning Multi-Layer Perceptron. The results highlight that the cost-sensitive Random Forest approach achieves higher accuracy, whereas the Multi-Layer Perceptron seems a good choice with huge amount of data or when features are unknown and hidden relationships need to be found. Hence with cases with a good and thoughtful features set, a standard machine learning approach can provide better results.

There are many directions for future works. An interesting variation is the reconstruction of conversational trees, where the issue is to identify the reply structure of the conversations inside a thread. Another more general development is studying the streaming version of the problem where identifying temporal thread discussions from a stream of messages rather than from a static pool of texts.

References

1. Jurczyk, P., Agichtein, E.: Discovering authorities in question answer communities by using link analysis. In: CIKM, Lisbon, Portugal, 6–10 November 2007, pp. 919–922 (2007)
2. Coussement, K., den Poel, D.V.: Improving customer complaint management by automatic email classification using linguistic style features as predictors. Decis. Support Syst. **44**, 870–882 (2008)
3. Glass, K., Colbaugh, R.: Toward emerging topic detection for business intelligence: Predictive analysis of meme' dynamics. CoRR abs/1012.5994 (2010)
4. Khan, F.M., Fisher, T.A., Shuler, L., Wu, T., Pottenger, W.M.: Mining chatroom conversations for social and semantic interactions. In: Technical report LU-CSE-02-011, Lehigh University (2002)
5. Hofmann, T.: Probabilistic latent semantic indexing. In: ACM SIGIR, pp. 50–57. ACM (1999)
6. Blei, D.M., Ng, A.Y., Jordan, M.I.: Latent Dirichlet allocation. J. Mach. Learn. Res. **3**, 993–1022 (2003)
7. Shen, D., Yang, Q., Sun, J., Chen, Z.: Thread detection in dynamic text message streams. In: SIGIR, Washington, USA, 6–11 August 2006, pp. 35–42 (2006)
8. Huang, J., Zhou, B., Wu, Q., Wang, X., Jia, Y.: Contextual correlation based thread detection in short text message streams. J. Intell. Inf. Syst. **38**, 449–464 (2012)
9. Adams, P.H., Martell, C.H.: Topic detection and extraction in chat. In: ICSC 2008, pp. 581–588 (2008)
10. Yeh, J.: Email thread reassembly using similarity matching. In: CEAS, 27–28 July 2006, Mountain View, California, USA (2006)
11. Domeniconi, G., Semertzidis, K., Lopez, V., Daly, E.M., Kotoulas, S., Moro, G.: A novel method for unsupervised and supervised conversational message thread detection. In: Proceedings of the 5th International Conference on Data Management Technologies and Applications, vol. 1, DATA, pp. 43–54 (2016)
12. Zhao, Q., Mitra, P.: Event detection and visualization for social text streams. In: ICWSM, Boulder, Colorado, USA, 26–28 March 2007
13. Lena, P., Domeniconi, G., Margara, L., Moro, G.: Gota: go term annotation of biomedical literature. BMC Bioinform. **16**, 346 (2015)

14. Ester, M., Kriegel, H., Sander, J., Xu, X.: A density-based algorithm for discovering clusters in large spatial databases with noise. In: KDD 1996, Portland, Oregon, USA, pp. 226–231 (1996)
15. Bouguettaya, A., Yu, Q., Liu, X., Zhou, X., Song, A.: Efficient agglomerative hierarchical clustering. Expert Syst. Appl. **42**, 2785–2797 (2015)
16. LeCun, Y., Bengio, Y., Hinton, G.: Deep learning. Nature **521**, 436–444 (2015)
17. Najafabadi, M.M., Villanustre, F., Khoshgoftaar, T.M., Seliya, N., Wald, R., Muharemagic, E.: Deep learning applications and challenges in big data analytics. J. Big Data **2**, 1 (2015)
18. Zhao, Q., Mitra, P., Chen, B.: Temporal and information flow based event detection from social text streams. In: AAAI, 22–26 July 2007, Vancouver, British Columbia, Canada, pp. 1501–1506 (2007)
19. Porter, M.F.: An algorithm for suffix stripping. Program **14**, 130–137 (1980)
20. Domeniconi, G., Moro, G., Pasolini, R., Sartori, C.: A comparison of term weighting schemes for text classification and sentiment analysis with a supervised variant of tf.idf. In: Data Management Technologies and Applications (DATA 2015), Revised Selected Papers, pp. 39–58, vol. 553. Springer, Heidelberg (2016)
21. Salton, G., Buckley, C.: Term-weighting approaches in automatic text retrieval. Inf. Process. Manage. **24**, 513–523 (1988)
22. Singhal, A.: Modern information retrieval: a brief overview. IEEE Data Eng. Bull. **24**, 35–43 (2001)
23. Manning, C.D., Raghavan, P., Schütze, H., et al.: Introduction to Information Retrieval, vol. 1. Cambridge University Press, Cambridge (2008)
24. Aumayr, E., Chan, J., Hayes, C.: Reconstruction of threaded conversations in online discussion forums. In: Weblogs and Social Media (2011)
25. Goodfellow, I., Bengio, Y., Courville, A.: Deep learning. Book in preparation for MIT Press (2016)
26. Sugomori, Y.: Java Deep Learning Essentials. Packt Publishing Ltd., Birmingham (2016)
27. Ulrich, J., Murray, G., Carenini, G.: A publicly available annotated corpus for supervised email summarization. In: AAAI08 EMAIL Workshop (2008)
28. Soboroff, I., de Vries, A.P., Craswell, N.: Overview of the TREC 2006 enterprise track. In: TREC, Gaithersburg, Maryland, USA, 14–17 November 2006 (2006)
29. Dehghani, M., Shakery, A., Asadpour, M., Koushkestani, A.: A learning approach for email conversation thread reconstruction. J. Inf. Sci. **39**, 846–863 (2013)
30. Erera, S., Carmel, D.: Conversation detection in email systems. In: Macdonald, C., Ounis, I., Plachouras, V., Ruthven, I., White, R.W. (eds.) ECIR 2008. LNCS, vol. 4956, pp. 498–505. Springer, Heidelberg (2008). doi:10.1007/978-3-540-78646-7_48
31. Wu, Y., Oard, D.W.: Indexing emails and email threads for retrieval. In: SIGIR, pp. 665–666 (2005)
32. Hall, M.A., Frank, E., Holmes, G., Pfahringer, B., Reutemann, P., Witten, I.H.: The WEKA data mining software: an update. SIGKDD Explor. **11**, 10–18 (2009)
33. Raschka, S.: Python Machine Learning. Packt Publishing, Birmingham (2015)
34. Wang, X., Xu, M., Zheng, N., Chen, M.: Email conversations reconstruction based on messages threading for multi-person. In: ETTANDGRS 2008, vol. 1, pp. 676–680 (2008)
35. Joshi, S., Contractor, D., Ng, K., Deshpande, P.M., Hampp, T.: Auto-grouping emails for faster e-discovery. PVLDB **4**, 1284–1294 (2011)
36. Wang, H., Wang, C., Zhai, C., Han, J.: Learning online discussion structures by conditional random fields. In: SIGIR 2011, Beijing, China, 25–29 July 2011, pp. 435–444 (2011)

Towards Efficient Multi-domain Data Processing

Johannes Luong[⊠], Dirk Habich, Thomas Kissinger, and Wolfgang Lehner

Database Technology Group, Technische Universität Dresden,
01062 Dresden, Germany
{johannes.luong,dirk.habich,thomas.kissinger,
wolfgang.Lehner}@tu-dresden.de

Abstract. Economy and research increasingly depend on the timely analysis of large datasets to guide decision making. Complex analysis often involve a rich variety of data types and special purpose processing models. We believe, the database system of the future will use compilation techniques to translate specialized and abstract high level programming models into scalable low level operations on efficient physical data formats. We currently envision optimized relational and linear algebra languages, a flexible data flow language(A language inspired by the programming models of popular data flow engines like Apache Spark (spark.apache.org) or Apache Flink (flink.apache.org).) and scaleable physical operators and formats for relational and array data types. In this article, we propose a database system architecture that is designed around these ideas and we introduce our prototypical implementation of that architecture.

1 Introduction

Due to increased processing power and improved analytical methods, corporations, government, and other organizations increasingly depend on large amounts of primary data to discover new information and to guide decision making. A growing variety of domain experts want to employ advanced data processing techniques as a flexible standard tool in their applications. This desire is in conflict with todays available processing systems which often rely on a programming interface that reflects low level systems requirements rather than high level domain specific abstractions. To use these systems, experts have to map high level application concepts onto domain independent operators and data formats that usually require in depth understanding to be used and composed efficiently.

We believe that the database community is in an excellent position to solve this problem. The requirements on modern processing systems are in many ways straightforward extensions to the ideas that have made classical relational databases widely successful. SQL and the relational algebra provide an established example for the decoupling of a high level application domain from its physical implementation. Users of relational databases formulate queries that are devoid of any low level system information. Database implementations automatically map these queries on adequate physical operators and data formats.

© Springer International Publishing AG 2017
C. Francalanci and M. Helfert (Eds.): DATA 2016, CCIS 737, pp. 47–64, 2017.
DOI: 10.1007/978-3-319-62911-7_3

Algebraic properties of the relational algebra introduce flexibility into the translation of queries that enables powerful optimization. In this way, relational databases have managed to marry an abstract high level programming model with excellent physical performance.

Modern relational databases form the backbone of a wide swath data intensive applications. Data warehouses and OLAP extensions make them an excellent match even for processing heavy analytical workloads. Nevertheless, it has been shown that there are important application domains, that are not served adequately by the classical relational model. On the one hand, the strict requirement for a static relational schema has turned out to be problematic in applications with a very dynamic development. On the other hand, an increasing number of important algorithms and domain concepts can not be mapped efficiently to the relational model. Advanced statistical and machine learning methods depend on iterative linear algebra processing and many internet related applications naturally match a rich graph model. Both models are not served well by the standard relational model.

In the first part of this article (Sect. 2) we introduce our vision of a data processing system that generalizes the idea of decoupled logical and physical representations to bring the ease of use and efficiency of relational databases to new important application domains. The core of this vision is an extensible domain specific compilation framework that can compile a set of domain specific high level languages into efficient physical workloads. Similar to the relational approach, the compilation framework exploits known algebraic properties of supported data types to efficiently map high level concepts to physical operators and formats. Besides algebraic languages, the compilation framework also supports a more traditional data processing model that depends on user defined functions and data parallel processing operators. This model is not as abstract and can not be optimized as well, but it provides additional flexibility that can be used to process unstructured data or to define custom operators that deviate from the algebraic language models. To emphasize our models support for multiple application domains, we call it the *multi-domain architecture* throughout the article.

The second part of the article (Sect. 3) is dedicated to the introduction of the *Eos* data processing environment. We develop *Eos* as a prototypical implementation of the multi domain architecture. The current version of *Eos* contains two low-level data formats, a set of corresponding physical processing operators, and a basic compilation framework. In the discussion of *Eos* we highlight the physical aspects of the system and provide a real world application scenario to motivate the selection of supported components. Subsequently, we introduce the *Eos* compilation framework and show how it can be used to cross compile domain specific languages into physical operator code.

In Sect. 4, we discuss our vision of domain specific optimization in some more detail and propose physical format transformations as a first inter domain optimization goal. In Sect. 5, we provide a brief review of related work on the topics of multi-domain processing and generative programming. Section 6 closes the article with a short discussion of our main findings. A short version of this article has been previously published in DATA 2016 [1].

2 The Multi Domain Architecture

We propose the *multi-domain architecture* as an architectural model for data processing engines. The goal of the architecture is to integrate efficient and easy to use processing of multiple data types in a single unified system. It uses a compilation framework to bridge the gap between high-level data processing languages and optimized physical data structures. The architecture comprises three layers: the language layer, the translation layer, and the physical layer (Fig. 1).

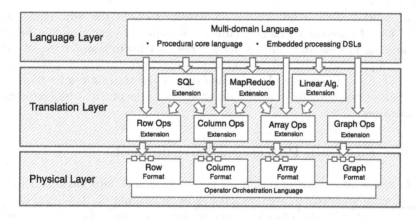

Fig. 1. Layers of the *multi-domain architecture* [1].

Language Layer. The *language layer* at the top of the architecture defines a *multi-domain* programming language for user defined processing tasks. The multi-domain language contains a set of embedded domain specific languages such as a SQL and a linear algebra dialect. It also provides a low-level operator language that closely mirrors the programming model of the *physical layer* and a slim procedural language that enables flexible program composition.

Users define their processing workloads using the language layer's core and embedded languages. They can use compiler optimized domain specific languages where possible and fall back to the full flexibility of lower level constructs when necessary. Additional languages and language elements can be added by extending the *translation layer* with additional compiler components.

Translation Layer. The translation layer's extensible compiler framework is the key to efficient processing in the multi-domain architecture. The compilation framework accepts language layer programs as input and translates them into optimized physical workloads. Where possible, the compilation framework uses algebraic laws of the input languages to facilitate code optimizations. The translation layer has access to statistical data that is collected by the physical layer. Algebraic laws and statistical data enable strong logical and physical optimization similar to query optimization in traditional DBMS.

Physical Layer. The *physical layer* at the bottom of the architecture provides efficient and scaleable data processing primitives. All higher-level constructs are mapped onto physical data structures eventually. Physical data formats such as tables or 2D arrays and their corresponding operators make different tradeoffs in flexibility and performance. This allows the architecture to achieve the best possible performance for each supported data type. The physical layer also has an orchestration language that is used to compose individual operator calls into larger workloads.

The multi-domain architecture is an attempt to carry the performance and ease of use of classical relational database systems into the world of flexible programming models and multiple data types. It tries to transfer the ideas of logical and physical relational optimization to new structures such as the linear algebra or graph processing. At the same time it celebrates low-level flexibility and expert intervention by providing direct access to the physical data structures and operators.

3   The *Eos* Data Processing Environment

The *Eos* data processing environment is a prototypical implementation of the *multi domain architecture*. We develop *Eos* in an ongoing effort, to explore and validate the architecture's practical implications and tradeoffs. *Eos* consists of two main components: *Eos Engine* and *Eos Script*. *Eos Engine* implements the *physical layer* and *Eos Script* implements both the *translation* and the *language layers*.

In the next section, we introduce the application scenario that guided the development of *Eos* up to the current version. The scenario will help us motivate the current selection of data formats and processing operators and it will provide a real-world background for code examples. Following, we will discuss *Eos Engine* in detail to gain a clear understanding of the low-level details of *Eos* data processing applications. The final section of part Sect. 3.3 is dedicated to the discussion of *Eos Script*, the compilation and language component of *Eos*.

3.1   An Eos Application Scenario

The development of the *Eos* data processing environment is guided by the goal to support domain experts in the analysis of a dataset in the size range of multiple tens of gigabyte [2]. The dataset encompasses high-frequency sensor data, that has been sampled from physical vibration sensors in wind mill installations, and low-frequency context information such as ambient temperature or wind speeds.

The sensor data is stored in the form of a large number of vibration frequency spectrums. Each spectrum corresponds to a measurement cycle that has been conducted in a wind mill at some point.

The wind mill domain experts have discovered that dangerous ice buildup on rotor blades correlates with certain high peaks in the vibration spectrums. Using their dataset of previously collected spectra and context information, the engineers try to build a regression model that predicts vibration peaks from context information. They hope to use the model to guide predictive counter measures such as shutting down the wind mill in certain weather conditions. In the current setup, the domain experts run a sequence of four octave scripts[1] to analyze the dataset.

1. *Data Selection:* The first script loads all spectra from csv files, matches the spectra with corresponding context information, and discards the spectra whose context data violates certain boundary conditions. The remaining spectra are saved to a file, using a format that can be efficiently loaded by subsequent scripts.
2. *Feature Extraction:* The second script loads the spectra that passed the environment filter of step one and searches the peak elements in each spectrum. The peak elements remain unmodified in the dataset but all non-peak elements are set to zero. The result of that transformation is stored on disk as well.
3. *Feature Reduction:* The third script loads the peak dataset and selects a single peak from each spectrum. The selection can be configured to return maximum peak value or to return the peak closest to a defined vibration frequency. The result of the feature selection is stored as a single column vector in a third dataset.
4. *Model Building:* The final script loads the previously selected peak frequencies and the context dataset. It builds a linear regression model on the selected peaks and their corresponding environment data. The result is a regression model that maps from environment variables to an expected peak frequency.

The domain experts choose octave because of its high level programming interface that facilitates the codification of domain logic. Unfortunately, the use of octave limits the performance of the workflow. Limited performance leads to long model development cycles that hinder the improvement of the modelling approach and its parameters. With the development of the *Eos* data processing environment we strive to replace octave in the spectra analysis workflow. Eventually, we hope to drastically improve performance, without reducing programmer productivity.

3.2   Eos Engine

Eos Engine is an in-memory data processing engine for multi-core shared-memory systems. Currently, it provides a table and a two dimensional array data

---

[1] https://www.gnu.org/software/octave/.

format as well as six customizable data processing operators. Besides, the engine also supports the linear algebra interfaces BLAS and LAPACK. *Eos Engine* is written in C++ and it relies on the parallelism primitives of the Intel® Threading Building Blocks[2] library. BLAS and LAPACK support is provided by the Intel® Math Kernel Library[3].

Eos Engine operates in batch processing mode. It maintains a job queue which is cleared in a first in first out manner. At any time, there is at most one active job in the system. Once started, a job runs from start to finish without interruption. Jobs are arbitrary C++ functions that execute in the engine's address space. An *Eos* job queries the system catalog to find data objects, uses special parallelized operators to process those objects, and eventually inserts result objects back into the catalog to make them available for inspection or subsequent jobs.

The goal of *Eos Engine* is to provide efficient, scaleable data processing to a single user. The user starts the engine, loads datasets from disk, runs analytical workloads in-memory, and eventually exports results back to disk. *Eos Engine* can be controlled with JSON encoded commands via HTTP or a Unix pipe. This flexible control mechanism makes it easy to integrate *Eos* as a component of a larger system or workflow.

Data Formats. Users can store data in the form of tables or two dimensional arrays. A table is defined as a sequence of columns of potentially varying types. For example, a table of type `table<uint62_t, string, double>` can store triples such as `(0, "pi", 3.1415)`. Tables offer an index based access interface, where users read and write table elements via $(row, column) \in \mathbb{N} \times \mathbb{N}$ indices. Internally, tables store data *by column*[4] to increase the efficiency of analytical workloads. These workloads often involve complete in-order column traversals and therefore benefit from fast column wise data access. Due to technical issues related to the static typing of tables and the C++ template mechanism, table columns have to be accessed using the free function `get<ColNr>(table<...>&)`. The table format is a natural fit for the context data of the wind mill use case. The context data contains signed and unsigned integers, floats, and even strings in one case. The table format allows to store the complete context in a single data object.

The two dimensional array format stores $(height \times width)$ values of a single data type. Similar to tables, matrix elements are accessed via $(row, column) \in \mathbb{N} \times \mathbb{N}$ indices. Behind the scenes, *Eos Engine* stores arrays in a contiguous chunk of memory in row-major order. In the current version, the major difference between the table and the array format is that the array format integrates seamlessly with the Intel® Math Kernel Library, which expects matrices defined in row- or column-major order. Besides that, the array format is essentially a (row, column) index based interface to a single column table. 2D arrays offer

---

[2] https://software.intel.com/en-us/intel-tbb.
[3] https://software.intel.com/en-us/intel-mkl.
[4] In contrast to *by record*.

```
1  using TableT = table<double>;
2  TableT in = catalog->get<TableT>("sample_table");
3
4  table_map_rows<TableT,TableT> select_positive;
5  TableT positive = select_positive(in,
6  [](const TableT& in, size_t start, size_t end, TableT& out) {
7      for (size_t row_idx = start; row_idx < end; ++row_idx) {
8        double row_val = get<0>(in)[row_idx];
9
10       if (row_val >= 0.0) {
11           out.insert(row_val);
12       }
13     }
14 });
```

Listing 1.1. Row filter with table_map_rows.

themselves as storage format for the spectra of the wind mill application. Spectra consist exclusively of double precision floating point values. All spectra can be stored row-wise in a single array object where they can be processed without further transformation by LAPACK's linear least square solvers to implement the final step of the wind mill scenario.

Processing Operators. *Eos Engine* encompasses six processing operators. Each of these operators implements an abstract, reusable data processing pattern. The abstract operators have to be parametrized with user defined functions (UDFs) to implement concrete application logic. *Eos'* operators are designed to enable implicit data parallelism, a form of parallel processing that is well adjusted to many of todays data analysis applications. The data parallel operators allow *Eos* to efficiently exploit hardware resources of common state of the art multi-core processors.

The most straightforward operator is table_map_rows<InT, OutT>. The operator divides its input table horizontally into segments of rows and processes each segment in parallel using a UDF. The UDF takes an input table of type InT, a begin and an end index, and an output table of type OutT as arguments. The begin and end arguments define the range of rows of the input table that the UDF is supposed to process. The UDF can insert an arbitrary number of result rows into the output table. Once all UDFs have returned, table_map_rows unions all output tables, which all have the same type OutT, into a single result table which is then returned as the result of the operator.

Listing 1.1 shows how table_map_rows can be used to filter negative numbers from a table. On line 4, the operator is constructed with input and output type and on line 5 it is invoked with the input table as first argument and the UDF as second argument. The UDF uses a for loop to iterate over each row of its table segment. Inside the loop, get<ColNr>(table<...>&) is used to retrieve the relevant column and operator[] to retrieve the current element of that column. If the value at that position is greater or equal zero (line 10) it is

```
1   using ArT = array2d<double>;
2   using TblT = table<double, double>;
3
4   ArT ar_in = catalog->get<ArT>("sample_array");
5   TblT tbl_in = catalog->get<TblT>("sample_table");
6
7   array_table_match_rows<ArT, TblT, ArT> select_rows;
8   ArT selected = select(ar_in, tbl_in,
9   [](ArT& ar, TblT& tbl, size_t begin, size_t end, ArT& out) {
10    for (auto row = begin; row < end; ++row) {
11      double mat_val = ar.at(row, 0);
12      double low_val = get<0>(tbl)[row];
13      double high_val = get<1>(tbl)[row];
14
15      if (low_val <= mat_val && mat_val < high_val) {
16          out.append(mat.row(row));
17      }
18    }
19  });
```

Listing 1.2. Array filter with array_table_match_rows.

inserted as new row into `out`. The `table_map_rows` operator is not used in the
wind mill application scenario. It has been included here because it offers the
most straightforward processing model of all operators. A prominent use case
for the operator would be the implementation of a SQL select clause.

The operator `array_table_match_rows<AInT, TInT, AOutT>` processes row segments using a UDF as well. In contrast to `table_map_rows`,
it accepts an array and a table as inputs and provides segments of both inputs
to its UDF simultaneously. The operator divides both inputs into row segments
of same height, pairs segments based on their segment index, and processes
each segment pair in parallel using a UDF. When all UDFs have returned, their
result arrays are concatenated in segment id order and the concatenated array is
returned as the result of the operator. `array_table_match_rows` can only be
applied, if both inputs have the same height. The UDF takes an input array of
type `AInT`, an input table of type `TInT`, segment begin and end indices, and an
output array of type `AOutT` as arguments. It can append an arbitrary number
of rows to the output array.

In Listing 1.2 `array_table_match_rows` is used to select rows of an array
whose values lie in a range defined by the input table. This use case matches the
first step of the wind mill application scenario where spectra are selected based
on conditions on context information. The UDF (starting at line 9) accepts only
those array rows, whose first element is greater or equal to the value in column 0
of the input table but smaller than the value in column 1 (line 15). In some sense,
`array_table_match_rows` is similar to a relational join operation because it
combines data from different inputs. On the other hand, joins use data dependent comparisons to find matching rows whereas `array_table_match_rows`
operates purely on row indices.

```
1  using ArrayT = array2d<double>;
2  ArrayT in = catalog->get<ArrayT>("sample_array");
3
4  array_stencil<0, 1, ArrayT, ArrayT> peak_finder;
5  ArrayT peaks = peak_finder(in,
6  [](ArrayT& in, vector<size_t>& rows, vector<size_t>& cols) {
7    double candidate = in.at(rows[1], cols[1]);
8    for (size_t c = cols[0]; c < cols[2]; ++c) {
9      double col_val = in.at(rows[1], c);
10     if (c != cols[1] && col_val >= candidate) {
11       return 0.0; // minimum value of the domain
12     }
13   }
14
15   return candidate;
16 });
```

Listing 1.3. Peak finder with array_stencil.

Both table_map_rows and array_table_match_rows apply their UDFs to completely independent input segments. In theory, this property allows them to scale nicely with the available hardware resources because sufficiently large data sets can always be split into enough segments to keep all processors busy.[5] The final operator who shares this beneficial property is array_stencil<Height, Width, InT, OutT>. It applies a UDF to each element of the input array in parallel and inserts the result at the same position into the output array. In addition to the element itself, the UDF can also access a rectangular neighbourhood around the element. The dimensions of the neighbourhood are defined by the operator's statically defined height and width parameters. The parameters define the thickness of the neighbourhood in horizontal and vertical direction. For example, a width of 1 signifies one neighbour to the left and to the right of the current element.

array_stencil's UDF accepts an input array of type InT, a row coordinate vector, and a column coordinate vector as arguments. Both coordinate vectors contain three elements: the minimum index, the index of the current element, and the maximum index of the respective direction. The UDF returns a single value that has to be compatible with the array type OutT as result. In Listing 1.3, array_stencil is used to find the peak row elements of an array dataset. A peak element is an element whose neighbouring elements have a strictly smaller value than the element itself. The stencil is configured to include a one element horizontal neighbourhood and zero vertical neighbours (line 4). The UDF simply iterates over each horizontal neighbour and checks if it is larger than the center element. If a larger element is found, the UDF returns zero, otherwise it returns the value of the center element to mark it as a peak value of the row. The sample code exactly matches the peak finding step of the application scenario.

---

[5] In practice, scaleability of data intensive workloads is often limited by memory bandwidth.

The final operator we are going to discuss is `array_reduce_rows<InT, OutT>`. This operator reduces the elements of each row of an array into a single element. The operator splits the input array by row and reduces the rows in parallel. A row is reduced by repeatedly replacing two consecutive row elements with their reduction value, until only a single value remains. The reduction operation is implemented by the UDF. The UDF accepts two row values as arguments and returns a single value as result. In contrast to the previous operators, the UDF calls can not all be applied in parallel. The results of a UDF call may need to be reduced again, enforcing partially sequential execution. In Listing 1.4, `array_reduce_rows` is used to compute the maximum element of each row of the input array. The simple UDF consists of a single return statement, which returns the larger of the two input values (line 7). The sample implements one possible peak selection strategy of step three of the application scenario. Alternative strategies such as selecting a peak that is closest to a given frequency can be accomplished with `array_reduce_rows` in a very similar fashion.

Eos Engine offers two additional operators: `table_reduce_rows<InT, OutT>` and `table_segmented_reduce_rows<PartitionCol, InT, OuT>`. The first reduces all rows of a table into a single result row. The second partitions the table on one of its columns and reduces each partition individually. They have been included to support a future SQL extension to *Eos* which would rely in these operators to implement aggregation functions and *group by* clauses. We do not give a detailed explanation of these operators because they closely resemble the previously discussed operators and because they are not used in the wind mill application scenario. The current set of operators is not meant to be complete in any capacity and we expect it to grow in the future to support additional requirements. The array format also supports BLAS and LAPACK operations. LAPACK least linear squares solvers are used in the final step of the application scenario, to compute the actual regression model. BLAS and LAPACK are well known packages that are reused unmodified by *Eos Engine*. For additional information on these libraries we refer to their respective documentations.

Jobs. *Eos Engine* is a job processing system, where jobs are processed in first in first out order. The engine uses the abstract C++ class *workload* to define the interface of a job. Subclasses of *workload* have to override the methods run()

```
1  using ArrayT = array2d<double_t>;
2  ArrayT arr = catalog->get<ArrayT>("sample_array");
3
4  array_reduce_cols<ArrayT, ArrayT> find_max;
5  ArrayT maxima = find_max(arr, 0,
6    [](double& a, double& b){
7      return a >= b ? a : b;
8    });
```

Listing 1.4. Reduction to maximum with array_reduce_cols.

and `getResult()`. When a job object reaches the front of the job queue, the engine removes it from the queue and calls its `run()` method. A job's `run()` method can use arbitrary C++ code to achieve its goals. On the other hand, ordinary processing jobs will usually just query the system catalog and call processing operators. A job is considered to have finished when its `run()` method returns. Some time after the completion of an operator, the engine might call its `getResult()` method to retrieve an arbitrary JSON object that describes the result of the job in some meaningful way. The result of `getResult()` is not used by the system, but simply returned to the user as the result of the operation.

New jobs can be loaded at runtime of the engine. To support this, the job has to be compiled as a shared library and provide an `extern "C" workload* makeWorkload()` hook function. The engine uses `dlopen()` and related linux functions to load shared library jobs into the *Eos engine*'s address space and calls `makeWorkload()` to retrieve a job instance which is then appended to the job queue.

Eos Engine is the *storage and processing layer* of the *Eos* data processing environment. It provides physical data formats, data processing operators, and a job execution system. In the following sections we will discuss *Eos Script*, which imlpements the *programming interface* and the *translation and optimization* layers of the multi domain architecture.

3.3   Eos Script

The data formats, processing operators, and the job system of *Eos Engine* are sufficient to implement any *Eos* application. Unfortunately, they are also low-level and require system programming skills to be used correctly and efficiently. The *Eos Script* component provides the tools to divorce application programming from system programming and to allow data analysts to work with domain specific concepts.

In the following sections we are going to discuss the *Eos Script* source to source compilation framework and show how it can be used to create data processing languages for the *Eos* environment. In this article, we lay out the fundamentals of the compilation process and develop a language that is still very close to the definition of *Eos Engine* jobs. Higher level languages and abstractions that depend on an advanced compilation process will remain future work. In Sect. 4 we provide a detailed discussion of our vision for optimized high-level data processing languages.

Operator Language. The operator language is the only currently available data processing language of *Eos Script*. The language closely reflects job definitions of the *Eos Engine*, introducing only minor abstractions. Despite its limited ambitions, the language touches most of the compilation framework's components. It will therefore provide a good basis to understand the details of the source to source compilation process.

```
1  val data = Array2D(DoubleT).from("filtered-spectra")
2  val peak_finder = Stencil(0, 1, DoubleT, DoubleT)
3  val peaks = peak_finder(data) {
4    (data, rows, cols) =>
5      val candidate = data(rows(1), cols(1))
6      for (col <- cols(0) to cols(2)) {
7        if (col != cols(1) && data(rows(1), col) >= candidate) {
8          return 0.0
9        }
10     }
11
12     return candidate
13 }
14 log("Saving peaks")
15 peaks.saveAs("peaks")
```

Listing 1.5. Peak finder with the operator language.

Listings 1.5 and 1.6 show peak detection and peak selection in the operator language. Despite some differences in syntax and naming, the code looks very similar to the previously discussed C++ implementations. However, instead of C++, the code is written in Scala. All data processing languages of *Eos Script* are in fact embedded domain specific languages (DSLs) for Scala. In contrast to other embedded DSLs, *Eos Script* DSLs have to be source to source compiled into C++. Because of this compilation, users can only use a carefully constructed subset of the complete Scala language. Only the language elements that provide a C++ translation rule can be used in *Eos Script*.

Source to Source Compilation. The core of *Eos Script* is its source to source compilation framework for embedded Scala DSLs. The basic components of the compilation are provided by the lightweight modular staging framework [3]. Lightweight modular staging (LMS) is an easily extensible framework that can transform Scala programs into a tree shaped compiler intermediate representation (IR). The framework defines IR traversal algorithms that can be extended to implement optimizations and code generation rules.

A Scala program that uses LMS is compiled into a code generator instead of a regular program. A LMS code generator is a standard java classfile that can be executed to generate the actual source code for the original input program. In the case of *Eos Script*, code generation creates a single C++ file that contains an *Eos*

```
1  val peaks = Array2D(DoubleT).from("peaks")
2  val peak_selection = ReduceColumn(DoubleT, DoubleT)
3  val selected_peaks = peak_selection(peaks) {
4    (a, b) => if (a >= b) a else b
5  }
6  log("Saving selected-peaks")
7  selected_peaks.saveAs("selected-peaks")
```

Listing 1.6. Peak selection with the operator language.

```
1   case class Log(x: Exp[Any]) extends Def[Unit]
2
3   def log(x: Exp[Any]): Exp[Unit] =
4     reflectEffect(Log(x))
5
6   override def emitNode(sym: Sym[Any], rhs: Def[Any]) =
7     rhs match {
8       case Log(x) => gen"""std::cout << $x << std::endl""";
9       case _ => super.emitNode(sym, rhs)
10    }
```

Listing 1.7. LMS IR element creation.

Engine job definition. Subsequently, the generated C++ has to be compiled into a shared library using a standard C++ compiler. The resulting shared library can be loaded and run by *Eos Engine*.

LMS does not rely on any kind of *"magic"* to transform a program into its intermediate representation. Instead, it provides methods to create and insert tree nodes that have to be called by DSL statements. Listing 1.7 shows how a DSL can implement a log(value) function that is compiled into a C++ cout << value << endl statement. The first line of the listing defines the type of the IR element that is created for each log call. All IR elements have to inherit from Def[T] where T is the type of the expression that is represented by the IR node.

The log(x: Exp[Any]) method definition on line 3 defines the syntax of the log DSL operation. log accepts the object that should be logged to the console as argument. Instead of a value of type Any it expects an IR node *reference* of type Exp[Any]. The implementation of log simply creates a Log IR tree node that stores the reference as a member variable. The call to reflectEffect(Log(x)) appends the Log IR node of type Def[Unit] to the list of children of the currently active block IR node and returns a node reference of type Exp[Unit]. Block IR nodes correspond to code blocks and LMS ensures that there is always an active block IR node.

The override def emitNode(...) definition on line 6 specifies the code generation rule for the Log IR node. emitNode is invoked by LMS' code generation IR traversal. The interface is not specific to any particular IR node so the implementation has to pattern match against the rhs to find out the type of the node. If the node is a Log, the method uses the gen string interpolator to emit some C++ code. gen resolves symbol references such as $x and then writes the code string to the currently open .cpp file.

Every language element of any data processing DSL has to be defined similar to the log method. The frameworks facilitates the creation of arbitrary complex compiled Scala DSLs by incrementally adding syntax definitions, node types, and code generation rules.

4   Optimization in a Multi-domain Environment

Compilers of general purpose languages usually can not reason about application specific properties of the applications that they translate. Therefore, they are also unable to perform application specific optimizations. In contrast, relational databases use a lot of domain specific knowledge and even dataset specific statistical information to optimize database queries. The optimizer knows that the order of selection predicates can be changed because the selection is commutative or that multiway joins can be parenthesized arbitrarily because of the associativity of joins.

The same principles can be applied to other data structures as well. Figure 2 shows two semantically equivalent tree representations of the linear algebra expression $M1 * M2 * v$. Tree 2(a) is the result of the default left to right evaluation of the matrix multiplication operator. Code generation based on that tree schedules the matrix-matrix multiplication ($res0 \leftarrow M1 * M2$) before the matrix-vector multiplication ($res1 \leftarrow res0*v$). On the other hand, linear-algebra domain knowledge tells us (a) that vector-matrix multiplications are cheaper than matrix-matrix multiplications, (b) that the result of a matrix-vector multiplication is another vector, and (c) that matrix multiplications are associative. Based on this knowledge, the optimizer should decide to perform the matrix-vector multiplication first, in order to replace the expensive matrix-matrix multiplication with a cheaper matrix-vector multiplication.

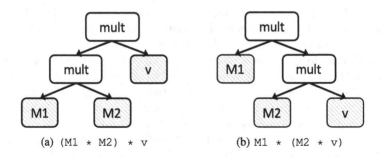

(a) (M1 * M2) * v   (b) M1 * (M2 * v)

Fig. 2. Simple Tree IR optimization [1].

The compiler framework of the *multi-domain architecture* is designed to compile an extensible but limited set of domain specific languages. Each domain extension can add its own specific compilation rules and thereby achieve results similar to classical relational optimization.

Besides optimizations that operate in a single isolated domain, such as the relational or the linear algebra, there is also potential for optimizations that cross the boundaries of domains. The topic of physical format transformations looks especially promising in that regard. Processing operators are data format specific and can only be applied to data objects of the correct type. It is easy to imagine a scenario where it would be convenient to apply an operator to a data object that has the wrong format. For example one might want to use a table column in a linear algebra expression that requires the use of the array format.

In general there are two ways to make this possible. One can either introduce a format conversion operator that takes a table column and returns an array or one provides a version of the relevant linear algebra operations that operates on tables. The first approach implies a small set of transformation operations that can be reused whenever a format mismatch occurs. The transformation operations can be targeted in optimization rules and, depending on the implementation, can be potentially merged into other operators. The conversion approach has the drawback that possibly expensive format transformations are applied independent of the cost of subsequent operations. In some cases, the performance penalty of an operation that is overloaded for a non optimal format might be much smaller than the added overhead of a transformation.

The second approach is based on the observation that many domain specific operations can be implemented on different formats, although with varying performance. This approach is beneficial in cases where the cost of a format transformation outweighs its benefit. The largest disadvantage of this approach is the necessity to provide multiple operator implementations.

We expect the ideal solution to be a combination of both approaches. For example, the compiler could use a cost model to decide whether to use a format transformation or an overloaded version of an operator. In general, the topic of format transformations is still an open research question of the architecture and we count on future work to find the best and most practical approach.

5 Related Work

In the following, we provide a brief discussion of related work on multi-domain processing and generative programming.

5.1 Multi-domain Processing

The need for multi-domain data management systems has been widely recognized. In this article, we propose to tightly integrate multiple storage formats and programing models into a single system. An alternative approach that has been discussed lately is the integration of several DBMS behind a data management middleware layer.

The BigDAWG *polystore* system [4] provides an example for this approach. The authors hide multiple *"of the shelf"* DBMS behind a central management layer, which defines a unified querying interface for all attached systems. The management layer accepts multi domain queries, splits these queries into parts that can be processed by one of the attached DBMS and sends the partial queries to the respective engines. The management layer also handles cases, where one DBMS depends on data that is currently stored in another DBMS and initiates the necessary data transfers.

In contrast to the *multi-domain architecture*, BigDAWG does not need to reimplement data management functionality and can instead reuse proven solutions. What is more, BigDAWG can incorporate DBMS that vary widely in important systems characteristics. BigDAWG could for example attach a relational in-memory DBMS and a classical file based DBMS at the same time and trade off processing speed versus durability guarantees on a per table basis.

On the other hand, data exchange between DBMS is a rather expensive operation as it depends on network communication. Frequent format changes are therefore much more feasible in the integrated single system approach, proposed by the *multi-domain architecture*. Furthermore, BigDAWG implies the administration of multiple separate DBMS, which increases the management cost compared to our integrated approach.

To summarize, the middleware approach provides greater flexibility with regard to non-functional properties but incurs a higher price for format transformations. In addition, administration of the system is more complex.

5.2 Generative Programming

Many big data applications repeatedly execute the same lines of code for millions or billions of data elements. Even expensive optimization becomes viable in that environment as their cost is amortized over time. This realization has sparked interest in runtime code compilation and compiler based optimizations. These two techniques trade additional one time compilation overhead for very efficient code that saves a couple of instructions for every data element.

Beckmann et al. introduce an embedded DSL for C++ that can be compiled, optimized, and executed at runtime of a host program [5]. They use the DSL to write image manipulation kernels that get optimized for specific transformation matrices and generate code that significantly outperforms standard solutions, given large enough image sizes. The primary efficiency gains of the generated code are based on removed indirections and runtime checks that are unavoidable in more general solutions. Newburn et al. follow a very similar approach and provide an embedded C++ DSL that is specifically targetted at data parallelism in multi-core systems [6]. They use code generation to specialize performance critical code passages to the specific runtime environment of their programs.

Another group [7] uses code generation, domain specific languages, and specific data access patterns to derive implicit parallelism in an approach that is very similar to the one described in this article. On the other hand they are not concerned with multi-domain integration and do not optimize for specialized data formats.

6   Discussion

This article introduces our vision for an easy to use data processing system that can map a variety of domain specific high level languages onto efficient physical operators. To increase flexibility and enable processing of unstructured data the system also supports lower level data flow processing. The system is based on the principle of decoupled logical and physical representations that has allowed relational databases to marry a high level domain specific programming model with an efficient physical implementation.

In the first part of the article, we have proposed the multi domain architecture as an abstract model for data processing engines that support our approach. At the top of the architecture, the language layer defines an extensible data processing language that contains both specialized high level as well as flexible lower level constructs. In the translation layer, we use a compiler framework to translate language layer programs into efficient physical workloads. The compiler incorporates domain knowledge which enables powerful domain specific optimization. At the bottom of the architecture, we propose to use a storage and processing engine that supports multiple optimized physical data formats.

In the second part we introduce *Eos*, our prototypical implementation of the multi domain architecture. Using a real world application scenario, we have discussed data formats and processing operators of the *Eos Engine* as well as the programming language and compilation model of *Eos Script*. Eos is still in early development and does not yet support many of the envisioned compiler features. Nevertheless we have touched all layers of the multi domain architecture and the successful implementation of the ice detection use case provides early validation of the approach.

In part three we have discussed some promises of domain specific compilation in greater detail. We have seen that strong logical optimization is not limited to the relational algebra and we hope to find interesting and beneficial domain specific optimizations for additional data structures in future work. We have also touched the topic of cross domain optimization and identified data format transformations as a promising are of research.

We close the article with a short discussion of related works in the fields of multi-domain processing and generative programming. The necessity to integrate different data processing models into an easy to use processing solution has received increasing attention lately and we believe to have found a very worthwhile research opportunity.

References

1. Luong, J., Habich, D., Kissinger, T., Lehner, W.: Architecture of a multi-domain processing and storage engine. In: Proceedings of the 5th International Conference on Data Management Technologies and Applications, DATA, vol. 1, pp. 189–194 (2016)
2. Aguilera, A., Grunzke, R., Habich, D., Luong, J., Schollbach, D., Markwardt, U., Garcke, J.: Advancing a gateway infrastructure for wind turbine data analysis. J Grid Comput. **14**(4), 499–514 (2016)
3. Rompf, T., Odersky, M.: Lightweight modular staging: a pragmatic approach to runtime code generation and compiled DSLS. ACM Sigplan Not. **46**, 127–136 (2010). ACM

4. Duggan, J., Elmore, A.J., Stonebraker, M., Balazinska, M., Howe, B., Kepner, J., Madden, S., Maier, D., Mattson, T., Zdonik, S.: The bigdawg polystore system. ACM SIGMOD Rec. **44**, 11–16 (2015)
5. Beckmann, O., Houghton, A., Mellor, M., Kelly, P.H.J.: Runtime code generation in C++ as a foundation for domain-specific optimisation. In: Lengauer, C., Batory, D., Consel, C., Odersky, M. (eds.) Domain-Specific Program Generation. LNCS, vol. 3016, pp. 291–306. Springer, Heidelberg (2004). doi:10.1007/978-3-540-25935-0_17
6. Newburn, C.J., So, B., Liu, Z., McCool, M., Ghuloum, A., Toit, S.D., Wang, Z.G., Du, Z.H., Chen, Y., Wu, G., et al.: Intel's array building blocks: a retargetable, dynamic compiler and embedded language. In: 2011 9th annual IEEE/ACM international symposium on Code generation and optimization (CGO), pp. 224–235. IEEE (2011)
7. Alexandrov, A., Kunft, A., Katsifodimos, A., Schüler, F., Thamsen, L., Kao, O., Herb, T., Markl, V.: Implicit parallelism through deep language embedding. In: Proceedings of the 2015 ACM SIGMOD International Conference on Management of Data, pp. 47–61. ACM (2015)

Approaching ETL Processes Specification Using a Pattern-Based Ontology

Bruno Oliveira[1] and Orlando Belo[2(✉)]

[1] CIICESI, School of Management and Technology, Porto Polytechnic,
Felgueiras, Portugal
bmo@estgf.ipp.pt
[2] ALGORITMI Centre, University of Minho, Braga, Portugal
obelo@di.uminho.pt

Abstract. The development of software projects is often based on the composition of components for creating new products and components through the promotion of reusable techniques. These pre-configured components are sometimes based on well-known and validated design-patterns describing abstract solutions for solving recurring problems. The data warehouse ETL development life cycle shares the main steps of most typical phases of any software process development. Considering that patterns have been broadly used in many software areas as a way to increase reliability, reduce development risks and enhance standards compliance, a pattern-oriented approach for the development of ETL systems can be achieve, providing a more flexible approach for ETL implementation. Appealing to an ontology specification, in this paper we present and discuss contextual data for describing ETL patterns based on their structural properties. The use of an ontology allows for the interpretation of ETL patterns by a computer and used posteriorly to rule its instantiation to physical models that can be executed using existing commercial tools.

Keywords: Data warehousing systems · ETL conceptual modelling · ETL patterns · Domain specific language and ontologies

1 Introduction

Ontologies are being used by many organizations to encode and share information across multiple systems, providing a way to electronic agents understand and use the information based on a solid formalism that can be shared. The need to reuse a particular domain knowledge is growing since it enhances better solutions and provides a better picture of a specific domain [1]. The struggle imposed by the global market affects business requirements in an unexpected way. Therefore, software design techniques should guaranty the quality and robustness of the software piece. The use of software patterns is a well-known reuse-based technique often applied in software developing on a lot of different domains [2]. The need to reuse components and share acquired knowledge across applications is crucial for reducing time and costs in software design and development, contributing to improve its final quality [3].

In the field of Data Warehouse Systems (DWS), ETL (Extract, Transform, and Load) processes represent the most important piece that supports a Business

© Springer International Publishing AG 2017
C. Francalanci and M. Helfert (Eds.): DATA 2016, CCIS 737, pp. 65–78, 2017.
DOI: 10.1007/978-3-319-62911-7_4

Intelligence system, consuming a large portion of the time and resources needed to its development. These processes are very particular and specific to each scenario where they are applied, since its main purpose is to integrate data from different data sources to target repositories, specifically built to support decision-making processes. The amount of data that is typically transformed associated to the specific data requirements and technology limitations that should be considered in its development places these systems in a special software domain [4]. All this contributes to increase the complexity of software development and maintenance. Additionally, there is still a lack of proposals and methodologies to support its development based on a conception phase that can successfully represents all operational stages with a simple notation, providing at the same time the necessary bridges for allowing its mapping to a physical model.

We propose a pattern-based approach designed to map typical operations - e.g. *Surrogate Key Pipelining* - SKP, *Slowly Changing Dimensions* - SCD and *Change Data Capture* - CDC techniques - to configurable components that can be adapted to specific scenarios. Based on previous works [5–7], and using the Web Ontology Language (OWL) [8], an ETL pattern based ontology is presented to support the necessary requirements and to describe each pattern configuration, enabling its mapping to physical models that can be posteriorly executed [9]. Basically, an intermediate layer is provided to separate technical knowledge typically used in commercial tools from the domain knowledge used by decision-makers [10]. Users should be capable to describe what to do without describing how it will be done. At the same time, users should be capable to extend and change patterns behaviour without affecting the whole system and by consequence, the remaining patterns.

Due the complexity of the knowledge involved, and the application of each pattern to specific contexts [11, 12], ETL processes can suffer from inconsistencies and misunderstandings related to communication problems that result in different meanings or architectural contradictions. Ontologies can be used to provide the contextual data to describe each pattern based on its structural properties [12]. The Web Ontology Language (OWL) [13] has been used to support ontologies development, describing a domain, its concepts and properties. Thus, after a brief exposure of some related work (Sect. 2), we describe our ontology approach to support ETL patterns, providing a specific taxonomy of the most used ETL techniques and the main components that support the configuration of each pattern (Sect. 3). Next, a set of necessary formalisms to create a pattern-based language and how to use them to generate physical models is presented (Sect. 4). Finally, we discuss the experiments done so far, analyzing results and presenting some conclusions and future work (Sect. 5).

2   Related Work

The development of more abstract models to support ETL processes development and their mapping to execution primitives is not new. Vassiliadis and Simitsis covered several aspects of ETL development in their research [14]. They cover the ETL conceptual modelling [15], its representation using logical views [16, 17], and its implementation using a specific ETL tool [18]. They based their approach on a specific notation that was used to support the specification of ETL activities. Skoutas [19] also explored the use of ontologies for ETL conceptual modelling, using a set of specific conceptual transformations based on ontology graphs. This approach is very interesting, however it is based in very specific constructs that are not integrated in a full framework that supports all the stages of an ETL development process using well-known technologies and tools. Thus, its implementation is still hard to accomplish using specific commercial tools.

More recently, Akkaoui [20] proposed a conceptual approach for ETL development based on known technologies such as BPMN (Business Process Model and Notation) and BPEL (Business Process Execution Language). Several mappings efforts were presented to support the mapping of BPMN models to BPEL executable models. This type of mappings is not easy and this approach suffers from very traditional problems already debated by research community [21]. Later, Akkaoui presents the BPMN4ETL metamodel [22], showing how conceptual primitives in BPMN can be mapped to physical models for commercial tools using specific templates.

In the field of ETL patterns, there is not much to refer. However, Köppen [23] presented a pattern-oriented approach to support ETL development, providing a general description for a set of patterns - e.g. aggregator, history and duplicate elimination patterns. This work focuses on important aspects defining patterns internal composition properties and the relationship between them. Patterns were presented only at conceptual level, lacking to support patterns instantiation for execution primitives. Some works also explored the use of UML (Unified Modeling Language) to develop ETL conceptual models [24, 25]. Later, Munoz [26] went further and presented the conversion of UML models to execution primitives. However, being a very strong language to describe system requirements, the UML is not so good to support execution of data based workflows such as the ETL processes. Our work distinguishes from the approaches presented so far since we followed a pattern approach based on well documented components that can be configured and used in different ETL development phases. Fine-grained tasks are encapsulated inside these components, resulting in a new ETL development level defined by the use of an upper abstraction layer that simplifies and carries the acquired knowledge between projects.

3   ETL Meta Model for Patterns Definition

Nowadays, sharing and reusing knowledge it is a crucial activity for software development. Several specific frameworks emerged to define a new kind of software programming that takes advantage of previous expertise and allow its reuse for new applications in different scenarios and domains. These frameworks are composed by collections of software patterns that represent a set of instructions or activities that can be configured and applied to more specific needs. In Web development, these frameworks are very used since they provide pre-established components that facilitate the creation of boring and error-prone tasks like website registration or login mechanisms. The use of software patterns allows for the identification of solutions for specific problems that can occur within a certain context. Thus a pattern catalogue that can be used in another projects in similar situations, can be used to develop and maintain software systems, contributing not only to higher software quality, but also to reduce time and costs required for its development can be reduced.

Concerning the specificities of the ETL environment, patterns can be characterized using a set of pre-established tasks that are grouped together based on a specific configuration related to the context in which they are used. Creating these reconfigurable components avoids the need to rewrite some of the most repetitive tasks typically used in these processes. Tasks such as surrogate key generation, lookup operations, data aggregation, data quality filters or slowly changing dimensions policies, are just some examples of some of the most usual procedures used in a DWS. Thus, instead of using repetitive tasks to solve the same problems, over and over again, conceptual models can be used to simplify ETL representation. Thus, ETL designers focus on more general requirements, leaving the complexity of its implementation to further steps. They only need to provide the configuration metadata to the conversion engine that will be responsible to generate the physical model. The ETL conceptual models can be created and used in posterior steps, enforcing the use of well-proven techniques, contributing to system quality and consistency. Despite the use of software patterns facilitate software development, some problems can emerge even with its use for ETL development. For example, the redundancy and different interpretations of the same concept can occur [11] for this design approach. Even under the same domain, different communities can have similar versions of same concept with slightly differences, leading to inconsistencies about how patterns can be used in different contexts. In an ETL environment different interpretations can occur using very traditional ETL procedures.

OWL, a language based on Web semantic technology, is often used to describe domain specific meta-models to represent properties and relationships between domain concepts (i.e. patterns). OWL is a W3C standard [27] that was developed to provide a simpler way to process and use semantic data across applications in the web. With OWL, classes or concepts can be described and arranged to form taxonomic hierarchies, properties describing the composition in terms of attributes of each concept and restrictions over the relationship between the concepts presented. The W3C specification document [28] describes OWL2 ontologies in three different syntactic categories: 1 – Entities such as classes, properties and individuals (class instances); 2 – Expressions that represent restrictions over the individuals; 3 – Axioms representing statements that can be asserted based on the domain description. Thus, the ETL patterns can be syntactically expressed using classes, data properties and object properties, providing the basic structure to support the development of a specific language to pattern instantiation. Figure 1 shows an excerpt of the breakdown among the different levels of the ETL patterns taxonomy proposed.

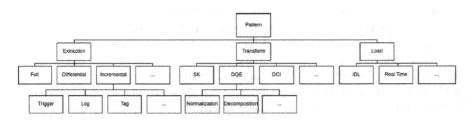

Fig. 1. The ETL patterns taxonomy.

The pattern concept is composed by a core that hides all structure to support each pattern operational requirements and the logic behind it, and the *Throwable* and *Log* components. These components encapsulate all logic related to exception and error handling for each pattern. The *Throwable* pattern uses the input configuration to handle error or exception scenarios through the application of specific recovery strategies for each pattern. For example, process errors that cause critical failure scenarios can be handled and rollback procedures can be used to preserve data in a consistent state. The *Log* pattern is responsible to store ETL events and its timeline to identify data lineage, bottlenecks and errors. Thus, the ETL process can be analysed and specific error trends can be found, revelling a need to handle and minimize them in source systems and eventually reduce ETL resources needed for subsequent loads. Log structures can differ in granularity level and scope. Its entries are triggered by conditions associated to each pattern or by more general conditions (such as process checkpoints). The following classes and slots are describe to support the ontology presented:

- Classes: The *Pattern* class contains the basic data properties for each pattern, e.g., the identification number (*Id*), *Name* and *Description*. The *PatternCore* encapsulates all logic behind each pattern operational requirements; The *Throwable* describes the exception/error handling processes and the *Log* describe the logging processes to track all pattern events.
- Object Properties: *HasCoreComponent*, *HasThrowableComponent* and *HasLog Component*, each one describing the relationship between individuals of each class involved and the main relationships between classes to support basic pattern structure. Each pattern can only has one *PatternCore* instance and zero or more *Throwable* and *Log* instances. This means that several *Throwable* and *Log* instances can be configured for the same pattern, allowing the description of several scenarios.

While *Pattern* class represents the most general concept used, the *Extraction*, *Transform* and *Load* classes represent the three types of patterns that are intrinsically associated to each typical phase of an ETL process. The *Extraction* class instances are used to extract data from data sources using a specific data object (e.g. a table or file), representing typical extraction data processes and algorithms applied over specific data structures. The *Extraction* class can be specialized in three more specific data extractors:

- *Full* extraction patterns that are used to extract all data from a specific data source without any criteria, i.e. all data currently available;
- *Differential* extraction patterns that are used to identify new data since the last successful extraction. For this data extraction type, all data from source and target repository is compared to identify new data.
- *Incremental* extraction patterns that are used to extract data from data sources since the last successful extraction but based on specific criteria and using specific CDC (Change Data Capture) techniques to identify and track the changed data in all the data warehouse data sources.

Due the much smaller data volume extracted, the incremental data extraction provides better performance when compared to differential extraction. However, sometimes is difficult to identify modified data due source systems access limitations. The trigger-based is one of the most used CDC techniques. This approach consists in the implementation of some triggering mechanisms inside the sources to capture insert, update and delete events. Triggers usually store the data they gather in specific audit tables that preserve a track of all the changes occurred in the source's tables, tagging all the records with some temporal and monitoring data. The log-based techniques are another example of incremental extraction that use the DBMS transaction log to identify new/modified records. Thus, analysing transaction log, it is possible to identify changes that occurred during some period in the operational system databases.

The *Transformation* class represents patterns that are used in ETL transformation phase for the application of a set of cleaning or conforming tasks [29] to align source data structures to the requirements of the target schema of a data warehouse. This class represents a large variety of procedures that are often applied in DWS, such as patterns responsible to apply the well-known policies related to SCD techniques, patterns for surrogate key generation, or patterns to support the conciliation and integration of data from many data sources. For example, a DQE pattern can be specialized to a *Normalization* class that represents the set of tasks needed whenever it is necessary to standardize or correct data according to a given set of mapping rules stored in mapping tables; and the *Decomposition* class that represents the set of tasks that support the decomposition of fields to extract meaning from all of its parts. These are just two examples of common DQE specializations typically used. With these classes, all the most frequent ETL patterns can be represented along with all its operational stages. Thus and using the ontology hierarchy to support ETL patterns meta-model, patterns can be changed or even new patterns can be added without compromising the whole pattern structure.

Finally, the *Load* class represents patterns that are used to load data to the target DW repository, representing efficient algorithms for data loading or index creation and maintenance for loading procedures. The *Intensive Data Loading* (*IDL*) subclass should load data to a target DW schema considering the model restrictions used. For example, based on multidimensional model approach [30], the dimension tables should be firstly loaded and only after the population of all dimensions, the fact table can be loaded.

After the taxonomy definition, the meta-model should be enriched to support the basic rules for the development of well-formed ETL patterns. For that, each class should be defined through the use of properties. For example, the *Extraction* class that represents all Extraction patterns is composed by some *Datatype* Properties such as *PatternId* and *PatternName* (inherited from *Pattern* class), and *Object* Properties such as *PeriodLiterals* that refers to extraction interval used (Hour, Daily, Month) (*oneOf* property) and the metadata related to the repository connection (*input* object: Data repository description class and associated fields: *Data* and *Field* class). The *Throwable* and *Log* components are not mandatory and multiple instances can be defined for the same pattern to represent different scenarios. Each subclass can also include additional or override properties. *Pattern* subclasses can override constraints using specific cardinality restrictions based on their own requirements. Sub properties are used for that, specializing their super properties. For example, the *Incremental* class uses a date type property to identify new or changed records. Each property should be described based on its cardinality, value, domain and range. The domain links a property to a class, while the range links a property to a class or data range. This allows the association between classes and data types, and provides a way to establish restrictions. The number of data repositories used for each pattern (both for input and output mapping configuration) should be ruled by each pattern specialization. For example, the DCI (*Data Conciliation and Integration*) pattern can use more than one data repository as input (using *subPropertyOf* axiom) due being responsible to integrate data extracted from several data sources related to the same data object and only one data repository as output.

The Fig. 2 shows a graph ontology summary of the main concepts that support pattern structure and configuration. The round corner rectangles are used to identify classes and rectangles used to identify object properties with domain and range properties. The three pattern components are identified along with the object properties to provide its configuration. The *SourceToLog*, *SourceToThrowable* and *SourceToTarget* object properties relate each pattern component configuration to specific mappings (*Mapping* class). The *Mapping* class describes the relationship between the input (*HasDataRepositoryInput* object property) and output (*HasDataRepositoryOutput* object property) *DataObject* instances representing data repositories that hold the data used for pattern configuration. This way, it is possible to establish relationships between data repositories attributes to enable data migration processes between them. These data repositories can have physical representation, e.g. databases and files, requiring a specific connection protocol or passed as stream of records.

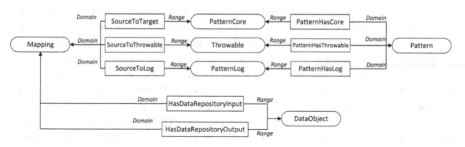

Fig. 2. Ontology graph representation

4 Pattern Language Specification

ETL systems are a very complex type of software that requires adjusting data with specific operational requirements in order to align their schema to new specific decision requirements, which makes ETL systems hard to develop and maintain. As mentioned before, several authors tried to simplify and minimize ETL systems development through the use of conceptual models that are used in early development phases. Currently, there is still a lack of semantics to support to express ETL systems and more importantly to provide the necessary mappings to execution properties, taking advantage of the work done previously in design phases. The majority of works presented till now supports ETL processes representation using very detailed tasks.

Thus, the generated models are composed by dozens of tasks without an automatic mapping to commercial tools. The use of detailed tasks such as joins or projections detail level affects process interpretation and are not very useful unless there is a specific tool to support their implementation.

With the pattern-based approach proposed, a new abstraction layer is proposed, simplifying and helping the ETL development process from conceptual phases to physical models that can be executed. For this particular task, we believe that commercial tools should be preferentially used, since they provide powerful and well-known frameworks that many IT professionals use. Therefore, we propose a specific configuration language that can be applied to each pattern presented by the ontology, covering its operational stages and providing a solid framework to enable its conversion to equivalent semantics used by ETL commercial tools. Using the Protégé-OWL API [31, 32], an ontology specification can be used and manipulated. Based on the concepts and properties presented, a specific generator was built to automatically generate a specific pattern configuration language, allowing for the configuration of each pattern using the ontology definition. The engine uses two important layers: the language construction rules (syntax) and the ontology data model. For the language specification, a set of specific statements and keywords were used to describe each language component.

The *USE* keyword is applied to identify the pattern path that should be followed based of the taxonomy presented (Fig. 1), followed by the pattern name. Top levels (*Pattern* class is the higher level) should be firstly defined and the character: . (dot) is used to traverse each hierarchy level from the middle levels to bottom levels. Next, and based on each *Pattern* class object properties, a set of blocks delimited by { } (braces) are defined. Inside each block, simple or composite assignments can be performed. For the general blocks (generated from *Pattern* class), simple assignments are formed based on data properties associated to *Pattern* class, while composite statements are generated based on the object properties. Each block can contain more than one occurrence based on the cardinality of the object properties associated to *Pattern* class. For example, the DCI pattern have several input blocks, each one related to the data repositories used (*SourceToTarget* object property) for data integration. The *OPTIONS* block is used to map the properties associated to the pattern class used and can be composed by single or composite statements (based on ontology definition). Table 1 resumes the main syntax used for the DSL (Domain-Specific language) proposed for pattern configuration.

Based on the ontology and the syntax rules presented, the configuration language can be automatically generated for each pattern, providing language flexibility. This approach guarantees that if the ontology is change, then the correspondent grammar rules will be consistent with the ontology definition. Figure 3 shows an example of the syntax rules applied to the language constructs and a correspondent example of its instantiation using a specific *Aggregator* pattern that applies a *sum* operation to the duration of telephone calls made by each customer. The *sum_duration* aggregator pattern (*Transform.Aggregator*) presents three main blocks derived from the object properties applied to the *Pattern* class. The *Source* describes input metadata, *Target* describes output metadata and *Fields* block describes the fields will be used as output to the target repository. These three blocks correspond to *hasInput*, *hasOutput* and *has-Fields* object properties, respectively. For input block, a CSV file was used for data

Table 1. Basic DSL constructs.

Language elements	Keywords	Ontology property	Example
Pattern selection	*PatternHierarchyPath.pattern_name*	ETL patterns hierarchy	Transformation.DQE
Atomic statements	Statement_name = expression	Class data property	Type = relational

Composite statements	[
	     statement_name_1 = expression_1
	     statement_name_n = expression_n
	]
		minCardinality value constraint	[
			 data=PickingSp
			 type=relational
			]

Main blocks	Block_name_1{} Block_name_n{}	Object properties with Pattern class	source{}

Multiple main blocks	Block_name{
	     [block_content],[(…)]
	 }
		maxCardinality value constraint	source{
			 [
			     data=PickingSp
			     type=relational
			]
			 },[]

Options	**OPTIONS{**
	     Atomic_Statement_name_1
	     Atomic_Statement_name_n
	 }
		Pattern data/object properties (*functionalProperty* or with *maxCardinality* = 1)	**OPTIONS{**
			 Function=SUM
			 Field=price
			     (…)
			 }

extraction based on delimiter ':' (a composite statement is used due the existence of a data property describing the delimiter rule for the *CSV* class), and the pattern output will store correspondent data into a specific relational table. Details such as database name or server were omitted since they can be configured in further steps. After fields identification (separated by comma), the keyword *OPTIONS* is used to specify each configuration parameter (derived from properties applied to *Aggregator* class) associated to Aggregator class: *Function* to identify the aggregation function applied, *FunctionField* to specify the field that should be used by the function, *RenameField* to apply the alias to the new field generated and the *GroupFields* used to specify the *group by* clause.

With the pattern-based approach presented in this paper, a new abstraction layer to develop ETL process is proposed. Patterns can be used to create conceptual model without focusing in very detailed tasks. However, to produce physical models based on conceptual primitives, two independent components should be provided: patterns configuration Meta data that is supported by the domain language provided, and workflow coordination data that describes the process flow. For demonstration purposes, the BPMN language was used to create ETL conceptual models. BPMN has proven in several works that is suitable to represent several workflow operational components of ETL systems both at conceptual and physical primitives [33, 34].

```
USE pattern_name
  block_name_1{
  ([)
   single_statement_name and
   [composite statement_name{...}],
    (]) (,) (...)
   }
  (...)
  OPTIONS{
      Pattern_options{...}
```

```
USE Transform.Aggregator
   'sum_duration'
   source{
    data=CDR_Calls.csv
    type=CSV{
     delimiter=':'
    }
   }
   target{
    data=calls
     type=relational
   }
   fields{
    DATETIME, CustomerId
   }
   OPTIONS{
    Function=SUM
    FunctionField=DURATION
    RenameField=TOTAL_DUR
    GroupFields=CustomerId
```

Fig. 3. Basic pattern configuration syntax and language example through the instantiation of an Aggregator pattern [7].

In recent works [9, 35] we proposed the use of BPMN as visual layer to support ETL conceptual models, representing patterns using BPMN elements. The experimental tool developed is responsible to interpret the configuration language and provide the generation of the physical model making it possible to be executed by commercial tools such as Kettle Pentaho [36]. Based on the ontology presented, a specific Meta model can be generated and used to support pattern instantiation and configuration. This feature allows for manipulating the ontology and, at the same time, provides the necessary contracts to control and implement the models that support pattern interpretation and manipulation.

The final step covers the generation of physical models using the architecture and philosophy followed by each commercial tool. A set of standard transformation skeletons was built to encapsulate the logic of the conversion process, providing the meanings to transform each pattern internal structure to a specific serialization format. To guarantee system flexibility and avoid the commercial tools proprietary formats, the Apache Velocity [37] template generator language was used to describe each component skeleton and build a specific and standard transformation template.

Conceptual design   **Logical design**   **Physical design**

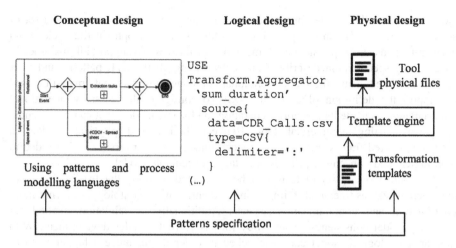

Fig. 4. ETL process development stages using a pattern-oriented approach [7].

Figure 4 summarizes all development process phases needed to support the physical representation of the ETL processes using patterns, from the ontology definition to the generation of physical model.

5   Conclusions and Future Work

Nowadays, companies need to adjust their business processes to meet new business demands, which implies the readjustment of their strategy and by consequence the processes used to store and process their data. The operational systems reflect these changes since operational data are stored in data schemas especially designed and built to serve particular business needs. The integration of new requirements in existing data schemas can lead to inconsistencies, because original schemas could not be prepared to support appropriately new business requirements. Due these reasons, the development of ETL processes is a sophisticated process that consumes a large amount of human and financial resources. ETL processes diverge from traditional data migration processes since they are used in a very specific domain area with specific architectural requirements that are already known.

The specificities of ETL systems have been studied and applied to several areas, contributing to the identification of common tasks and solutions in order to solve them. The documented SCD and data conciliation and integration policies are just two examples of already discussed techniques applied in the majority of ETL projects. However, this knowledge is carry on between projects as technical documentation, describing guidelines and good practices that should be applied. Instead, we believe that this knowledge can be parametrized and encapsulated using container of tasks, grouped together according to a specific purpose. Thus, we believe that the ETL early development stages can be simplified, replacing the traditional large number of tasks and operators by simpler composite tasks that encapsulate pattern logic.

Additionally, the knowledge and best practices can be put in practice using a set of software patterns that can be applied to the entire ETL development life cycle: from conceptual phase to its physical implementation primitives. To support all this process, an ontology specification describing and categorizing all the ETL patterns and their rules is proposed. Thus, the main operational components of each pattern can be used to support the definition of a specific DSL to configure all necessary operational requirements to enable its posterior generation to physical models that can be executed using an existing commercial tool. Between logical and physical models, a template-oriented framework is used to encapsulate all the conversion logic needed to map logical and conceptual primitives to target executable model.

As future work, a set of tests will be conducted to study the feasibility of our approach as well as to extend it, improving and enriching the ontology, covering more coordination and communication aspects. Additionally, a validation method for checking model consistency using Alloy [38] is currently under development. With this, an ETL logical model can be checked in order to guarantee a higher level of consistency and correctness before the generation of their physical representation.

References

1. Gruber, T.R.: A translation approach to portable ontology specifications. Knowl. Acquis. **5**, 199–220 (1993)
2. Gamma, E., Helm, R., Johnson, R.E., Vlissides, J.: Design patterns: elements of reusable object-oriented software. Design. **206**, 395 (1995)
3. Alexander, C., Ishikawa, S., Silverstein, M.: A Pattern Language: Towns, Buildings, Construction. Oxford University Press, Oxford (1977)
4. Weske, M., van der Aalst, W., Verbeek, H.: Advances in business process management. Data Knowl. Eng. **50**, 1–8 (2004)
5. Oliveira, B., Belo, O.: BPMN Patterns for ETL conceptual modelling and validation. In: 20th International Symposium on Methodologies for Intelligent Systems (ISMIS 2012), Macau, 4–7 December 2012
6. Oliveira, B., Santos, V., Belo, O.: Pattern-based ETL conceptual modelling. In: Cuzzocrea, A., Maabout, S. (eds.) MEDI 2013. LNCS, vol. 8216, pp. 237–248. Springer, Heidelberg (2013). doi:10.1007/978-3-642-41366-7_20
7. Oliveira, B., Belo, O.: An ontology for describing ETL patterns behavior. In: Proceedings of 5th International Conference on Data Management Technologies and Applications (DATA 2016), Lisboa, Portugal, 24–26 July 2016
8. McGuinness, D.L., van Harmelen, F.: OWL Web Ontology Language Overview (2004)
9. Oliveira, B., Belo, O.: A domain-specific language for ETL patterns specification in data warehousing systems. In: Pereira, F., Machado, P., Costa, E., Cardoso, A. (eds.) EPIA 2015. LNCS (LNAI), vol. 9273, pp. 597–602. Springer, Cham (2015). doi:10.1007/978-3-319-23485-4_60
10. McGuinness, D.L., Wright, J.R.: Conceptual modelling for configuration: a description logic-based approach. Artif. Intell. Eng. Des. Anal. Manuf. **12**, 333–344 (1998)
11. Dietrich, J., Elgar, C.: Towards a web of patterns. Web Semant. Sci. Serv. Agents World Wide Web **5**, 108–116 (2007)

12. Noy, N., McGuinness, D.: Ontology development 101, A guide to creating your first ontology. Development. **32**, 1–25 (2001)
13. Antoniou, G., Van Harmelen, F.: OWL web ontology language. Handb. Ontol. Inf. Syst. **2007**, 157–160 (2004)
14. Vassiliadis, P., Simitsis, A., Georgantas, P., Terrovitis, M.: A framework for the design of ETL scenarios. In: Eder, J., Missikoff, M. (eds.) CAiSE 2003. LNCS, vol. 2681, pp. 520–535. Springer, Heidelberg (2003). doi:10.1007/3-540-45017-3_35
15. Vassiliadis, P., Simitsis, A., Skiadopoulos, S., Conceptual modeling for ETL processes. In: Proceedings of the 5th ACM International Workshop on Data Warehousing and OLAP, DOLAP 2002, pp. 1–25 (2002)
16. Vassiliadis, P., Simitsis, A., Skiadopoulos, S.: On the logical modeling of ETL processes. Science **80**, 782–786 (2002)
17. Simitsis, A., Vassiliadis, P.: A method for the mapping of conceptual designs to logical blueprints for ETL processes. Decis. Support Syst. **45**, 22–40 (2008)
18. Vassiliadis, P., Vagena, Z., Skiadopoulos, S., Karayannidis, N., Sellis, T.: ARKTOS: a tool for data cleaning and transformation in data warehouse environments. Bull. IEEE Comput. Soc. Tech. Comm. Data Eng. 1–7 (2000)
19. Skoutas, D., Simitsis, A.: Ontology-based conceptual design of ETL processes for both structured and semi-structured data. Int. J. Semant. Web Inf. Syst. **3**, 1–24 (2000)
20. El Akkaoui, Z., Zimanyi, E.: Defining ETL worfklows using BPMN and BPEL. In: Proceedings of the ACM Twelfth International Workshop on Data Warehousing and OLAP, DOLAP 2009, pp. 41–48 (2009)
21. White, S.A., Corp, I.B.M.: Using BPMN to model a BPEL process. Business **3**, 1–18 (2005)
22. El Akkaoui, Z., Zimànyi, E., Mazón, J.-N., Trujillo, J.: A model-driven framework for ETL process development. In: Proceedings of the ACM 14th International Workshop on Data Warehousing and OLAP, DOLAP, pp. 45–52 (2011)
23. Köppen, V., Brüggemann, B., Berendt, B.: Designing data integration: the ETL pattern approach. Eur. J. Inform. Prof. **XII**, 49–55 (2011)
24. Luján-Mora, S., Trujillo, J., Song, I.-Y.: A UML profile for multidimensional modeling in data warehouses. Data Knowl. Eng. **59**, 725–769 (2006)
25. Muñoz, L., Mazón, J.-N., Pardillo, J., Trujillo, J.: Modelling ETL processes of data warehouses with UML activity diagrams. In: Meersman, R., Tari, Z., Herrero, P. (eds.) OTM 2008. LNCS, vol. 5333, pp. 44–53. Springer, Heidelberg (2008). doi:10.1007/978-3-540-88875-8_21
26. Muñoz, L., Mazón, J.-N., Trujillo, J.: Automatic generation of ETL processes from conceptual models. In: Proceedings of the ACM Twelfth International Workshop on Data Warehousing and OLAP, pp. 33–40. ACM, New York (2009)
27. W3.org, Semantic Web - W3C. http://www.w3.org/standards/semanticweb/
28. Motik, B., Patel-Schneider, P.F., Parsia, B., Bock, C., Fokoue, A., Haase, P., Hoekstra, R., Horrocks, I., Ruttenberg, A., Sattler, U., Smith, M.: OWL 2 Web Ontology Language - Structural Specification and Functional-Style Syntax, 2nd edn. Online, pp. 1–133 (2012)
29. Rahm, E., Do, H.: Data cleaning: Problems and current approaches. IEEE Data Eng. Bull. **23**, 3–13 (2000)
30. Kimball, R., Ross, M.: The Data Warehouse Toolkit: The Complete Guide to Dimensional Modeling. Wiley, Hoboken (2002)
31. Protégé, The Protégé Ontology Editor (2011)
32. Horridge, M.: protégé-owl api. http://protege.stanford.edu/plugins/owl/api/
33. Akkaoui, Z., Mazón, J.-N., Vaisman, A., Zimányi, E.: BPMN-based conceptual modeling of ETL processes. In: Cuzzocrea, A., Dayal, U. (eds.) DaWaK 2012. LNCS, vol. 7448, pp. 1–14. Springer, Heidelberg (2012). doi:10.1007/978-3-642-32584-7_1

34. Oliveira, B., Santos, V., Gomes, C., Marques, R., Belo, O.: Conceptual-physical bridging - from BPMN models to physical implementations on Kettle. In: CEUR Workshop Proceedings, pp. 55–59 (2015)
35. Oliveira, B., Belo, O., Cuzzocrea, A.: A pattern-oriented approach for supporting ETL conceptual modelling and its YAWL-based implementation. In: 3rd International Conference on Data Management Technologies and Applications, DATA 2014, pp. 408–415 (2014)
36. Bouman, R., Van Dongen, J.: Pentaho® Solutions: Business Intelligence and Data Warehousing with Pentaho and MySQL® (2009)
37. Gradecki, J.D., Cole, J.: Mastering Apache Velocity - Java Open Source library (2003)
38. Jackson, D.: Software Abstractions: Logic, Language, and Analysis. MIT Press, Cambridge (2012)

Topic-Aware Visual Citation Tracing via Enhanced Term Weighting for Efficient Literature Retrieval

Youbing Zhao, Hui Wei[✉], Shaopeng Wu, Farzad Parvinzamir,
Zhikun Deng, Xia Zhao, Nikolaos Ersotelos, Feng Dong,
Gordon Clapworthy, and Enjie Liu

University of Bedfordshire, Luton LU1 3JU, UK
{youbing.zhao,hui.wei,shaopeng.wu,farzad.parvinzamir,
zhikun.deng,xia.zhao,nikolaos.ersotelos,feng.dong,
gordon.clapworthy,enjie.liu}@beds.ac.uk

Abstract. Efficient retrieval of scientific literature related to a certain topic plays a key role in research work. While little has been done on topic-enabled citation filtering in traditional citation tracing, this paper presents visual citation tracing of scientific papers with document topics taken into consideration. Improved term selection and weighting are employed for mining the most relevant citations. A variation of the TF-IDF scheme, which uses external domain resources as references is proposed to calculate the term weighting in a particular domain. Moreover document weight is also incorporated in the calculation of term weight from a group of citations. A simple hierarchical word weighting method is also presented to handle keyword phrases. A visual interface is designed and implemented to interactively present the citation tracks in chord diagram and Sankey diagram.

Keywords: Text mining · Citation tracing · Data management · Ontology · Term weighting · TF-IDF · Visualization

1 Introduction

Topic based retrieval of scientific and research documents can offer significant assistance to researchers by providing them the most relevant documents within their research interest. While citation analysis has attracted much attention in research communities, little has been done to incorporate topic-based document analysis with citation tracing.

Effective citation analysis of large corpuses of scientific and research documents involves a wide spectrum of techniques, including document indexing for the creation of numeric representations of documents; ranking of key scientific terms; and weighted representations of the documents, etc. Term selection and weighting are used to identify the most relevant terms and assign a numeric value to each term to indicate the contribution of the term to its document. Citation relationships captured in time can not only facilitate literature retrieval but also indicate the evolution of research topics over years.

© Springer International Publishing AG 2017
C. Francalanci and M. Helfert (Eds.): DATA 2016, CCIS 737, pp. 79–101, 2017.
DOI: 10.1007/978-3-319-62911-7_5

This paper presents a detailed work on topic-enhanced visual citation tracing of large corpuses of scientific literature based on our previous work [1, 2].

At the pre-processing stage, text mining is used to extract citation relations (namely the reference list) and metadata (title, year, authors, etc.) obtained from raw PDF files. The extracted information is then stored in the document repository. Standard terms from a document are collected with their occurrence after lemmatization and Stop Words removal.

The data management is implemented by following a NoSQL scheme in order to address scalability. We have studied characters of different types of NoSQL data repositories which are employed for retrieving information. CouchDB was selected because of its on-the-fly document transformation. A semantic repository, the Sesame RDF, was used to describe key scientific terms and their synonyms in the CG field. An external resource MAS keyword API (MAS API) is used as the input data to create the ontologies.

The citation relationships between the documents in the repository are analysed and stored using a graph repository, enabling quick citation path retrieval. From a pair or a group of related citations, an improved term-weighting scheme, which selects important terms according to their relevance to the cited documents, is employed. It takes into account the popularity of the scientific terms in the relevant year, as well as their occurrence in the entire SIGGRAPH corpus. Terms appearing in higher ranked documents should be given higher weights.

The citation relationships are finally visualized using a directed graph controlled by a user-specified citation track length. The graph shows all paths that satisfy the restriction imposed by the path length. The weighted terms are shown in the graph in descending order.

In summary, our contributions are as follows:

- an approach for the management of large scale corpuses of scientific documents that work seamlessly with the underlying text mining framework to support efficient document retrieval based on topics and citation relationships
- a term weighting scheme allowing for the ranking of key scientific terms over years at both the document level and corpus level
- a visualization method to display citation relationships between the scientific documents together with weighted scientific terms.

The rest of the paper is organized as follows. Section 2 provides an overview of related works, Sect. 3 describes the design requirements and Sect. 4 describes our term weighting method. Section 5 presents our approach to visualization. Section 6 discusses the implementation and Sect. 7 concludes our work.

2 Related Work

Organizing, management, analyzing, and exploring massive text information has been a highly interested research area for decades. Researchers from a variety of domains have devised methods to categorize and mine the large corpuses of available scientific literature. Information science is a field particularly devoted to developing data analysis methods for this goal. Text mining, document analysis and document visualization are closely related to the research work in this direction. And database technologies are very helpful for the implementation of a concrete document mining system. Consequently, the related work is organized in three sections: data management, text mining and document visualization.

2.1 Data Management

NoSQL databases are increasingly used in big data and real-time web applications and it has also been concluded that graph databases are more efficient in traversing relationships [3]. Kivikangas & Ishizuka introduced a semantic representation format Concept Description Language (CDL) [4]. They store semantic data presented by CDL in Neo4j [5] graph database and utilize semantic relationships to improve query performance.

Compared to only one or two data repositories used in the data layer support of most applications, four NoSQL repositories are designed and employed in our work to facilitate high efficient indexing and queries.

2.2 Text Mining

Term selection and term weighting (TW) are important processing phases in text mining and have been investigated for many years.

A term-weighting scheme can affect not only text classification, but also other text mining tasks, such as sentiment analysis, cross-domain classification and novelty mining [6]. A classic term-weighting scheme introduced in [7] is based on three assumptions: 1. the multiple occurrence of a term in a document is related to the content of the document itself; 2. terms uncommon throughout a collection better discriminate the content of the document; 3. long documents are not more important than short ones, so normalize the length of documents.

By applying sorted term-weighting at a document level important terms can be revealed from repeated or redundant terms, thus enabling quick extraction of the most useful information [8].

TF-IDF has long been proposed for text mining and document topic analysis [9]. It is the product of two statistics, term frequency (TF) and inverse document frequency (IDF). With TF-IDF, the topics of a document can be calculated based on term weighting. There are many variants for determining the exact values of term frequency and inverse document frequency.

Supervised term weighting and a number of "supervised variants" of TF-IDF weighting are proposed in [7] for image recognition applications that involve supervised leaning, such as text filtering and text categorization.

Domeniconi et al. [10] proposed a supervised variant of the TF-IDF scheme, based on computing the usual IDF factor without taking documents of the category to be recognized into account. The idea is to avoid decreasing the weight of terms included in documents of the same category, so that words appearing in several documents of the same category are not undercounted. Another variant they proposed is based on relevance frequency, considering occurrences of words within the category itself.

Li et al. [11] proposed a cross-domain method extracting sentiment and topic lexicons without counting labelled data in the interested domain but counting labelled data in another related domain.

Another cross-domain approach [12] creates explicit representations of topic categories, which can be used for comparison of document similarity. The category representations are iteratively refined by selecting the most similar target documents. Further, [6] compared and discussed the impact of TW on the evaluation measures, and recommended the best TW function for both document and sentence-level novelty mining.

None of these works uses citing relations as a factor in term weighting.

2.3   Text and Document Visualization

Text and document visualization and visual analysis has been a field of high interest in the visualization community for decades. [13–15] present very good surveys on text and document visualization and visual analysis. As our work is on citation analysis and visualization, we focus specially on visualization and visual analysis of document relationships. A systematic review of citation analysis and visualization has been presented in [16].

CiteSpace II [17] uses node-link diagrams to visualize co-citation with a focus on the interplay between research fronts and their intellectual bases. They apply time slicing, thresholding, modelling, pruning, merging and mapping methods to prune a dense network. Zhang et al. [18] organize paper references in a tree structure and Citevis [19] presents citation links in a matrix-based visualization. Jigsaw [20] correlate documents and other entities based on metadata and content in visualisation and PivotPaths [21] visually integrates citation data with other document metadata to form an explorable network. CitNetExplorer [22] delineates research fields and help literature reviewing based on analysis of a very large citation network with node-link diagram visualisation. CiteRiver [23] facilitates user-steered aggregation of citations and supports the exploration of the dataset over time and enable users to analyze citation patterns and trends.

Our work enhances citation analysis with topic analysis based on an improved term weighting model. Interactive visualisation of citations in chord diagrams and Sankey diagrams offers seamless integration of document topic information.

3   Design Goals and Requirements

3.1   Design Requirements

The motivation of our work is to assist scientific literature search by professionals with more efficient citation tracing which incorporates topic analysis. Based on the motivation we have the following design goals:

1. List citing and cited papers of a given publication

Citations are bi-directional and document management should support retrieval and tracing of citing papers as well as cited papers. This retrieval should return a list of citing or cited papers of a given document.

2. Retrieve and trace citations from a given publication

The goal requires retrieval and tracing of all direct and indirect cited papers of a given document. With this function, all literature related to a given document can be found. This retrieval or tracing should return all direct and indirect citing or cited papers of a given document.

3. Topic analysis of citing papers and cited references

The citing papers of a given document may have different topics and some may be more related to the user's interests. This function evaluates the topics of the citations, categorizes them into different groups and evaluate the relevance to the given document.

The cited references of a given document may also come from different domains and have different topics. It is also beneficial to researcher's work if those references can be grouped in topics and evaluated against the user's interests. This function evaluates the topics of the references and categorizes them into different groups.

4. Detect the longest citation track

From the citations within the dataset, the longest citation track can be identified, which helps to retrieve the early work and all the related work pertaining to a given publication. It is also crucial in studying the provenance and trend of a research work.

5. Effective interactive visualisation

Visualisation is indispensable for presenting complex and big data sets to facilitate human understanding. Document contents and relationships are targets of visualisation applications. In our work, we are interested in visualizing the citation tracks as well as document topics. Well-designed visualisation improves the efficiency and effectiveness of citation retrieval and topic analysis.

3.2   The Dataset

The work proposed in this paper is a generalized method for topic based scientific literature retrieval and analysis. However, to evaluate the method effectively we need domain experts. The dataset used in this paper are papers from ACM SIGGRAPH [24] conference proceedings. ACM SIGGRAPH is a large research community on computer graphics and it is the top conference on computer graphics. The annual ACM SIGGRAPH conference has a track of more than 30 years to provide a large number of documents for time-varying analysis. In this paper we use papers of ACM SIGGRAPH conference proceedings from year 2002 to 2014 as the data source for document and citation analysis.

4 Data Management

In the implementation we define 4 logical data entities: Citation, Corpus, Reference and Keyword. A Citation is a published paper that is stored in our system in full text and PDF. A set of Citations published in the same year form a Corpus. A Reference is a cited paper in the reference list of a Citation (a paper). A Keyword of a citation is a CG keyword that appears at least once in a paper.

As mentioned earlier, we use ACM SIGGRAPH conference proceedings from 2002–2014, which include 1228 publications during 13 years. Corpuses are organised by years, which naturally introduces a time factor for topics. This natural corpus is used as the logic corpus.

The raw resource of a Citation (a paper) is a semi-structured PDF file generated from a domain specific template, ACM Proceedings Template in our work, which facilitates the implementation of text mining to extract META data of each citation.

For a Reference in the reference list of a Citation, we extract the title, year and authors as its identity. This reference could either be a citation that already exists in the system, or it could be a publication outside of the SIGGRAPH scope. In this paper, we assume that SIGGRAPH has already covered a history of major topics in CG and only references that can be matched to citations within our system are considered for text mining. The outgoing references are stored but not processed.

Although the keyword list section in an academic paper represents the author's selection of keywords, it alone can hardly reflect topics effectively in most cases as authors may use different phrases to represent the same concept, such as "3D"/"three dimensional", "level of detail"/"LOD", and so on. To resolve this ambiguity, an ontology is introduced to represent explicit specification of the shared concept.

To handle the complexity that resides in the data, four types of data storage are employed for efficient data management and information retrieval including a semantic repository, an index and search repository, a document repository, and a graph repository. These repositories are designed to work in coordination to effectively store and index data with reliability and efficiency for scientific text mining. The implementation of these repositories are introduced in the following subsections.

4.1 Semantic Repository

The standard keyword list we use for the shared concept is fetched from the MAS API which provides a keyword function to return keyword objects in a variety of fields. For the "computer" area, it covers 24 fields in total, including "computer graphics", "computer vision", "machine learning", "artificial intelligence" etc. For the "computer graphics" field which is the interest of this paper, a collection of 13670 keywords are provided by the MAS API [25].

Each CG keyword in the CG field is denoted by an ontology graph model with nodes and edges, represented in RDF (Resource Description Framework) as an instance with "rdf:type" of CG. The synonyms are described by the "owl:sameAs" predicate. The outcome of this work is that each keyword in a citation can be mapped to a node with type CG in the semantic repository. We choose Sesame [26] as our RDF repository as it provides APIs for RDF creation, parse, storing, inferencing and query. It can also be connected to the Semantic annotation tool GATE [27] used for META data extraction. The frequency of each keyword can be calculated from the "GATE ontology, Gazzetter producer" output.

4.2 Document Repository

The implementation of data management takes advantage of NoSQL in order to address scalability. The performance and characteristic of different types of NoSQL data repositories have been investigated before CouchDB [28] was selected due to its on-the-fly document transformation.

As a NoSQL document repository designed for web application, CouchDB allows files be treated as attachments of a document. By passing the document id, attachments of a document can be accessed easily. Since CouchDB treats each record as a document without considering its properties, the database can accommodate a large number of documents. Each document in the database is assigned a docType value which is used to distinguish document types from corpus, citation, keyword frequency and doc references. The docType property plays the role of a table in relational databases that holds a structured format with collection of related data. For documents, a virtual table of data structures is created for this schema-less repository.

Some benefits of a CouchDB document repository are:

- It provides a design document "View" to sort documents by the key of a view as in any relational database. Furthermore, values emitted from the view can not only be fetched from the database directly but also be calculated from functions written in Javascript.
- It provides validation functions in design documents with the property name validate_doc_update. To be valid, each document has to satisfy all these functions for creation and update. Consequently, the data structure of documents in this schema-less database can be guaranteed.
- It provides the Reduce function to reduce a list to a single value, which is useful for creating a summary of a data group by data aggregation.

4.3 Graph Repository

The citation relationship can be obtained from the document repository by querying citing documents and cited documents. However, detecting similar citing or cited documents of two given papers requires massive deep queries that are not optimised. To favor efficient relationship query the graph database which is tailored to traverse of relationships is used in our system. In a graph repository, a relation is described by a path and it is much more efficient and convenient to perform path queries in a dedicated graph repository. In our work, the graph repository Neo4j [29] is selected as the graph database to store citation relationships as well as relationships between citation and keywords. In Neo4j, each entity is represented as a node identified by a LABEL.

We define a document as a node entity and "cite" as a relationship that directs from one citation source node to the target node. The graph repository loads data from the document repository to create relationships of reference (citation) and usage (keywords) at the data processing stage. The following are representative tasks the graph repository undertakes (Fig. 2):

A.   Navigate and retrieve cited citations from a given citation.
B.   Citing/cited papers from a given paper.
C.   Detect similar citing citations.
D.   Detect similar cited citations.
E.   Detect the longest path.

As mentioned earlier, when processing the reference list of a paper, an outgoing reference which is not managed by our system it will be left unchanged. Alternatively for a reference points to an existing document in the document repository, the two objects are merged into one object in the graph repository. From the citing and cited documents in the document repository, citation tracks which reflect the evolution of interest in the context of CG can be built. With the 1228 citations from SIGGRAPH, the citation tracks are too complex to be presented in a node-link diagram. Figure 1 only presents a simplified directed graph of the data model in the graph repository with only the documents and their cited documents where the longest citation track has a length of 8 and covers the whole 13 year span, as shown in Fig. 3.

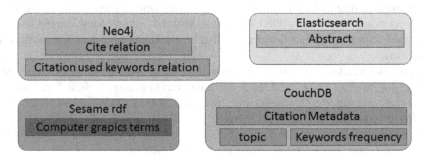

Fig. 1. NoSql data repositories employed.

4.4   Search Repository

Views in a document repository are the primary tool for querying the CouchDB documents. A View function accepts parameters and returns emit [key, value] pairs as a result. Query by user defined keywords is a main method to search for related papers. However, if user-defined keywords are used as parameters to query a view in CouchDB, they need to be included by a key emitted from this view. Unfortunately from a predefined virtual table structure, it is not possible to predict the properties that will be searched by the user. Based on this observation, the Elasticsearch engine [30] is employed to provide a document-oriented, full-text search via a RESTful API.

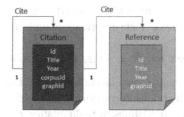

Fig. 2. Relations of citing & cited publications.

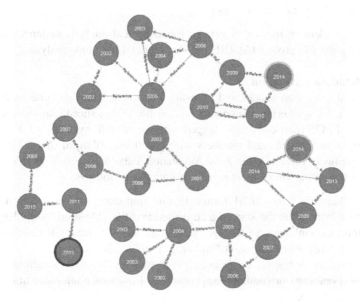

Fig. 3. Citation tracks in the graph repository.

The CouchDB document repository stores brief description of the corpus information, title, author, year and the full text part of a paper as an attachment. The search function is provided by an Elasticsearch plugin called Mapper Attachments Type [31]. With the brief description, the searched papers returned from the search engine contain all necessary information for a list presentation and no further information retrieval from the document repository is needed.

5 Text and Document Analysis

5.1 Data Collection

As most of the conference papers are in PDF format, the meta information of documents has to be extracted from the original PDF documents. With the aid of Apache PDFBox [32], an open source Java tool for working with PDF documents, the original

PDF files are first converted into plain text files and then organized into corpuses based on properties from the extracted meta information such as organization, conference, year, etc.

The data collection phase has three stages listed as follows:

- extraction of meta information such as authors, title, abstract sentences, doi, which can be used to provide the brief information as well as the unique ID for each paper.
- extraction of references and citations to build citation chains for each paper among the publications.
- extraction of standard key terms from each paper and calculation of their frequency to estimate the topics of the paper.

The data collection method is general for many text analysis systems and is also employed by the EC project CARRE [33] for medical literature analysis.

5.1.1 Metadata Extraction

The Metadata extraction is achieved by using a text processing pipeline supported by the GATE Text Engineering Framework. A text file is first sliced into sentences, then tokenized and POS (part-of-speech) tagged with the ANNIE system [34], followed by recognizing person names and numbers with gazetteers. With all this information, a series of grammar rules (JAPE: Java Annotation Pattern Engine [35] rules) is then applied for the extraction of meaningful information as follows:

1. Define "Macro"s from ACM format to find important markers, such as "ACM Reference Format" at the start of a converted text file; "Abstract" tag, "Keywords" tag, "Introduction" tag, "DOI" tag, "year" tag, "Author" tag, "CR Categories" tag, etc. Make JAPE rules to output "author", "year", "title" tags.
2. References are extracted by searching for two consecutive occurrence of {authorTag}{yearTag} or {authorsTag}{yearTag} from which reference title, year and authors can be extracted.
3. CR categories are extracted by searching for abstract sentences, located between the "Abstract" tag and the "CR Categories" tag.
4. Keywords are extracted by searching for keywords list sentences starting from "the Keywords" tag, ending at the "Introduction" tag.

The above are general rules applying to ACM publications. However, modifications are needed for publications over different periods. If applying these rules successfully, each paper is mapped to one metadata and is stored in the document repository.

Figure 4 shows an example of metadata extraction, including titleTag, authorTag, yearTag doiTag etc. extracted with multi tags. Citations are extracted with citing title, author and year tags. Keywords in cgTag are also extracted from the text to help keyword frequency calculation.

5.1.2 Keyword Extraction

The Microsoft Academic Search (MAS) API [36] allows developers to build applications by leveraging the data and functions of MAS. They supply a keyword function that represents keyword objects in a number of fields including Biology, Chemistry,

ACM Reference Format
Zhu, J., Lee, Y., Efros, A. 2014. AverageExplorer: Interactive Exploration and Alignment of Visual Data
Collections. ACM Trans. Graph. 33, 4, Article 160 (July 2014), 11 pages. DOI = 10.1145/2601097.2601145

Abstract
This paper proposes an interactive framework that allows a user
to rapidly explore and visualize a large image collection using the
medium of average images. Average images have been gaining
popularity as means of artistic expression and data visualization, but
the creation of compelling examples is a surprisingly laborious and
manual process. Our interactive, real-time system provides a way to
summarize large amounts of visual data by weighted average(s) of
an image collection, with the weights reflecting user-indicated importance. The aim is to capture not just the mean of the distribution,
but a set of modes discovered via interactive exploration. We pose
this exploration in terms of a user interactively "editing" the average
image using various types of strokes, brushes and warps, similar to
a normal image editor, with each user interaction providing a new
constraint to update the average. New weighted averages can be
spawned and edited either individually or jointly. Together, these
tools allow the user to simultaneously perform two fundamental
operations on visual data: user-guided clustering and user-guided
alignment, within the same framework. We show that our system is
useful for various computer vision and graphics applications.
CR Categories: I.3.8 [Computer Graphics]: Applications—;
Keywords: big visual data, average image, data exploration
Links: DL PDF

References
AGARWALA, A., DONTCHEVA, M., AGRAWALA, M., DRUCKER,
S., COLBURN, A., CURLESS, B., SALESIN, D., AND COHEN,
160:10   •   J.-Y. Zhu et al.
ACM Transactions on Graphics, Vol. 33, No. 4, Article 160, Publication Da
ANGELOVA, A., ABU-MOSTAFAM, Y., AND PERONA, P. 2005.
Pruning training sets for learning of object categories. In CVPR.
BALCAN, M.-F., AND BLUM, A. 2008. Clustering with interactive
feedback. In Algorithmic Learning Theory, Springer, 316–328.
BELHUMEUR, P. N., JACOBS, D. W., KRIEGMAN, D. J., AND
KUMAR, N. 2011. Localizing parts of faces using a consensus
of exemplars. In CVPR.
BERG, T., AND BERG, A. 2009. Finding iconic images. In 2nd
Workshop on Internet Vision.
BERG, T. L., BERG, A. C., AND SHIH, J. 2010. Automatic at-
tribute discovery and characterization from noisy web data. In
ECCV.

ACMFormatTag
AbstractSentenceTag
AbstractTag
AuthorTag
AuthorYearTag
CategoriesTag
DoiTag
IntroductionTag
KeywordsSentenceTag
KeywordsTag
Lookup
PaperAuthorTag
PaperTitleTag
PaperYearTag
RefAuthorTag
RefAuthorsTag
RefStartTag
RefTag
RefTitleTag
RefYearTag
Sentence
SentenceCgTag
SpaceToken
Split
Token
cgTag

Fig. 4. Metadata extracted and highlighted.

Engineering Mathematics, Physics, Computer Science, etc. In the computer science category, domains like "Algorithms & Theory, Artificial Intelligence, Computer Vision, Data Mining, Databases, Graphics" are described separately. We target our research in the "Computer>Graphics" domain, from which we collected 13,670 key-words. In a scientific publication, the usage of a group of keywords may reflect its topic. These 13K keywords are employed as standard terms to match phrases used in a computer graphics paper. In citation topic analysis, these 13K keywords are used for every citation to calculate the keyword frequencies. These keywords need to be stored it in a suitable repository that can be easily mapped with standard words in the Computer Graphics (CG) citation context.

The keywords fetched from MAS API contain a variety of terms for the same concept, such as "three dimensional", "3D" and "three-dimensional", which will be treated as different terms with naive machine processing. To fix this problem, an ontology is introduced into our system to share this conceptualization [8, 9]. These "3D" synonyms should be treated as same "type" in the ontology with multiple "same as" links. We specially define predicate "rdf:type" and "owl:sameAs" for this purpose in Sesame [10] RDF repository and convert all these standard terms into Resource Description Framework (RDF) triples. By building up this "OWLIM-LITE" repository with "Owl-max" ruleset, this repository is connected to the GATE Gazetteer as the ontology source.

5.2   Term Weighting

As mentioned earlier a standard keyword list which contains more than 13K key words is used for topic analysis of each citation. However, most citations in CG field use less than 100 standard keywords out of 13,670. Term weighting is used to evaluate the topics of the each citation, where the occurrence or frequency of each citation related keywords is calculated. Frequent appearance indicates more importance of the corresponding keyword. Term weighting helps a variety of text mining tasks including text classification, topics extraction and sentence analysis. In this paper it is used to further help topic analysis of the citations in a large corpus.

The keyword part in MAS API supplies the name of the keyword along with two other important properties: publication count, which indicates the number of publications of each keyword, and citation count, which presents the total number of citations among all publications using this keyword.

Table 1 shows the top 10 keywords in CG, sorted by citation numbers. There are 13670 keywords in this field in total. One may notice that, some of the top keywords used in the field of computer graphics also appear in the top 10 keyword list of other fields. For example the keyword "real time" appears in Computer Vision (12839 keywords) as well, as some domains have similar research topics to others (Xinyi 2015). The more documents a term appears in, the less effectiveness it is in distinguishing document topics. In text mining, the Inverse Document Frequency (IDF) is used with the term frequency to reflect the importance of a word to a document in a collection or corpus. Equation 1 shows one common definition of IDF. With the introduction of IDF, terms appearing frequently in the corpus are expected to have less importance. This can help to filter out the more common terms.

$$idf = \log\left(1 + \frac{N}{n_t}\right) \tag{1}$$

We use 4 different levels of characteristic terms to calculate the term frequencies and inverse document frequencies: field level, citation level in CG, year level in CG, and hierarchical topic names as described in the following sections.

Table 1. Top 10 keywords sorted by citation counts.

Keywords	#Publication	#Citation
Computer graphics	4729	99608
Real time	4208	68950
Three dimensional	2131	46419
Text mapping	1010	34038
Geometric model	1028	30263
Volume rendering	1418	29171
Ray tracing	1195	29061
Virtual environment	1904	28844
Virtual reality	2342	27932
Level of detail	1146	27846

5.2.1 Field Term Weighting

The field term weighting is dedicated term weighting of characteristic terms for each field. Keywords from MAS API in 24 fields of the computer domain are retrieved and treated as 24 documents. In the keyword corpus of a domain, $D = \{d_1, d_2, ..$ $d_j, \ldots d_{|D|}\}$, each file contains the keywords with occurrence of publication count or citation count as shown in Table 1. In the CG domain, the document contains 13670 keywords. A document d_j thus can be represented as an n-dimensional word vector $w_i = \{w_{1j}, w_{2j}, \ldots, w_{nj}\}$ with each word mapped to a weight factor w_i in the document.

We assume the data fetched from MAS API is counted from a large corpus of related field citations. Hence the citation count property of each keyword in a field document can be regarded as the occurrence value of this keyword in a field of this corpus.

The weight factor of keywords in each field document is calculated based on TF-IDF, the combination of the raw frequency and inverse document frequency (IDF), as shown in Eq. 1:

$$F_{w_j}(d_i, w_j) = Tf * \log\left(1 + \frac{N}{n_t}\right) \tag{2}$$

where N is the number of the total fields (here 24), and n_t is the occurrence of a keyword in other fields. Tf is the citation count of each keyword in Table 1.

The outcome of this is that in CG, each keyword is assigned a weight indicating its importance in CG compared to other fields. This result is used as a global weighting value and mapped to a local weighting result such as Citation Term Weighting and Year Term Weighting.

5.2.2 Citation Term Weighting

For citations of different fields, each citation emphasizes different topics even if they have similar frequent terms. Occurrence of a term is highly dependent on the context. Field term weighting introduced in Sect. 5.2.1 also indicates the relevance of a term to a field compared to other fields. In this section, citation terms that are different from other citations in the same corpus are identified by calculation of the local IDF L_{idf} for each citation keyword with the following equation:

$$L_{idf} = Tf * \log\left(1 + \frac{N}{n_t}\right) \tag{3}$$

where N is the citation number of our corpus (1228) and n_t is occurrence of this keyword in other citations.

By using the MapReduce function provided by the document repository, a summary of occurrence of keywords in citations can be obtained as each keyword is associated to a citation id. For each citation, a keyword only has one frequency value. If this value is mapped to 1, the reduced result will be the occurrence (n_t) of the keyword in all citation documents, as calculated by the following equation.

$$C_{w_j}(d_j, w_j) = Tf * F_{w_j}(d_j, w_j) * L_{idf} \tag{4}$$

where Tf is the citation related term frequency in the document repository, which identifies important keywords for this citation in the field of CG.

5.2.3 Year Term Weighting

Research topics change with time. When studying the trend and changing of research topics, a year-based topic model is commonly used, which in turn requires keyword weighting of citations calculated at a yearly basis. Moreover keyword weight and citation weight interact with each other. A citation with a higher weight implies that the keywords in this citation should be weighted higher than those in citations with lower weights. In other words, document weighting contributes to term weighting when calculating term weighting in a group of citations.

A straightforward way to assign a score to a citation is to find the citation counts. As the citation relationship A → B for each citation has been stored in the graph repository, querying the number of A will give the citation counts of B, which is referred as "Score(d_j)" in the paper. Finally the term weight which considers the influence of a citation is calculated as follows:

$$Rank(d_j, w_j) = Score(d_j) * C_{w_j}(d_j, w_j) \tag{5}$$

Assuming the number of documents in a year is n_{year}, the term rank over a year can be calculated by Eq. 6.

$$Rank(w_j, year) = \sum_{j=1}^{n_{year}} Rank(d_j, w_j) \tag{6}$$

5.2.4 Hierarchical Word Weighting

It is not rare that keywords are phrases in which the occurrences of certain component words are more meaningful to the keyword. Examples in the rendering topic of the CG field are "image based rendering", "real time rendering", "non-photorealistic rendering", etc. In this paper these keywords are named as "hierarchical words". To calculate the importance of a hierarchical word more accurately in its field, we designed an alternative to the TF-IDF.

Our method of hierarchical word weight calculation comprises three steps. In the first step, a phrase keyword is treated as a group of individual words which contributes to its own keyword equally with the score of the citation count of that keyword. The score of a word which appears in multiple keywords is calculated as the sum of the corresponding keyword weight in all related citations. The hierarchical word weight is calculated for each field of the 24 keyword fields of MAS API. For each field, the score represents the term frequency (TF) in one document.

In the second step the occurrence of the hierarchical words in all the 24 fields are calculated by summing the TF value over all 24 fields, which is named as TotalTF in this paper.

Finally the importance of each hierarchical word is calculated based on improved TF-IDF with the focus of improving term weight over document weight as shown in Eq. 7. The method is based observations that words occurring less in a field should be less important than those occurring more.

$$R_{w_i} = \sqrt{TF} * idf^2 \tag{7}$$

where

$$idf = \log(1 + \frac{TF}{a + (TotalTF - TF)}) \tag{8}$$

This improvement increases the global factor and reduces the local factor, and leads to higher accuracy in this context. The constant a is added to the denominator to avoid a zero $TotalTF - TF$ which means the term appears in no fields but the CG, and has therefore an important indicative to the CG field.

5.2.5   Citation Distance

Citation relationships are strong connections between documents. In the citation diagram, the width of a citation link can represent the strength. As mentioned earlier, each citation is associated with a keyword frequency list acquired from the document repository. Percentile weighting of each keyword in a citation indicates the importance of a keyword relative to the citation. Citations share more common keywords in higher percentile values have higher cosine similarity, i.e. smaller citation distance.

5.3   Longest Citation Track Query

By querying the graph repository the longest citation track in the graph database can be found. However, the data size of a graph may be large as each link are described by URIs of the start node, end node and the relationships. Direct returning a graph via the Restful API may result in low performance. Therefore, to obtain further properties from the URIs, more sophisticated query designs are needed.

```
"data" : [ [ {
    "directions" : [ "->" ],
    "start" : "http://localhost:7474/db/data/node/226",
    "nodes" : [ "http://localhost:7474/db/data/node/226", "http://localhost:7474/db/data/node/117" ],
    "length" : 1,
    "relationships" : [ "http://localhost:7474/db/data/relationship/223" ],
    "end" : "http://localhost:7474/db/data/node/117"
  } ],
  -
]
```

Fig. 5. Node-link data structure for citation track query.

Our solution to this problem is divide-and-conquer and progressive retrieval which transforms the graph query task into a series of sub-queries: retrieve the start node ID list first, followed by search paths starting from those start nodes. As the longest path is 8 in our database, the start node ID, all the relationships can be represented in a directed node-link graph data structure, as shown in Fig. 5. Traversal of a citation track can then be converted to a series of directed node pairs.

In Fig. 6, the Cypher query language in Neo4j provides a variety of functions that help to query data at low cost, such as multi-match, filter, id, etc. For two tracks at length 8, 14 pairs of id list are returned from the graph database, which can be used to generate the citation tracks directly.

```
MATCH p= (a: Citation) <-[r: Refer * 8 ..] - (b: Citation)

where a.id='b7ca85bb_0e57_4f38_b800_772215579bd2'

WITH [x in NODES (p) | ID (x)] AS vp1, [y in NODES (p) | ID (y)] AS vp2

MATCH (n1: Citation) <-[r1]-(n2: Citation)

WHERE ID (n1) in vp1 and ID (n2) in vp2

RETURN distinct n1.id, n2.id
```

Fig. 6. Convert a track to a list in Cypher.

6   Visualisation

We design an interactive tool to represent citation relations as a direction graph. A node represents one citation, and a link represents the relationship between two citations. An in-direction link of a node means it is cited by another citation; an out-direction link means it is citing other citation.

6.1   Visualisation of Citations

Based on reference information extracted from the PDF files, a large number of citations are generated from the SIGGRAPH proceedings of 13 years. Citations within SIGGRAPH can be tracked, analysed and visualized. As the initial attempt, we visualize the citations with in D3.js [37] chord diagrams where documents are placed on the circle and the chords represent citations. Citations of all visible documents can be presented in a single chord diagram. The citation graph can be treated as many trees if

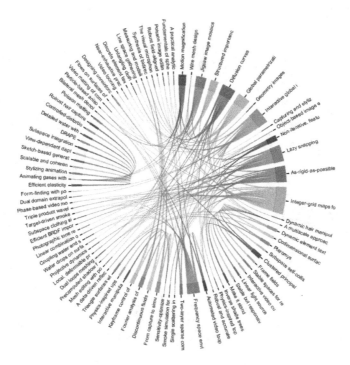

Fig. 7. Chord diagram visualization of cited tracks at length 6 from year 2002 to 2014.

we treat the start nodes as root nodes of their reference trees. In the chord diagram, filtering is applied to visualize trees within the specified depth. The specific citation can be highlighted when the user hovers the mouse over it.

Figure 7 presents a visualization of citations across 13 years with filtering of the trace path greater than 6. Papers linked with short arcs are normally published in recent years and papers with longer arcs are published normally in earlier years. The graph database provides much more efficient query of these citation tracings than other types of database. The longer the arc is, the more citation traces with a length greater than 6 are associated to the paper. Due to limited screen space, all paper titles are truncated to 20 characters. Figure 8 presents the highlighting of citations of the given document by mouse hovering on a paper title and the hiding of other citations. Figure 9 presents citations of papers published in a single year 2013. Those papers which cite the papers in 2013 must be published in year 2014 and later. From Fig. 9 it can be easily recognized that a paper in brown named "Globally optimal dir" has the highest citation count.

6.2 Visualisation of Citation Tracks

The disadvantage of the chord diagram is that it can only show pairwise relationships but cannot show citation tracks effectively. A Sankey diagram [38] is a better choice for visualising relationship routes. The advantage of the Sankey diagram is that it shows

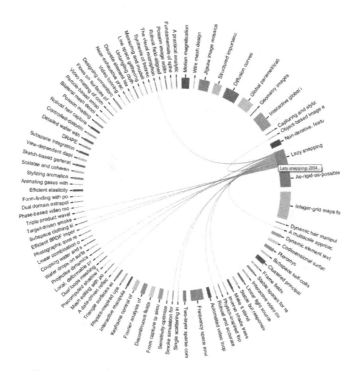

Fig. 8. Mouse hovering to highlight citations between 2002 and 2014 of a given document. (Color figure online)

the citation tracks in a much more clear and understandable way than the chord diagram.

Therefore, Sankey diagram is used to visualise the citation tracks, as shown in Fig. 10. The node height is proportional to the weight sum of its outgoing links. Nodes with same colour are documents published in the same year. The weights of the links are presented by link width.

When a mouse hovers on a node, a semi-transparent tooltip appears showing details of the citation. Only the citation ID, year and title of META data are shown to reduce overlaps. The information is acquired by querying the citation ID of the node in the document database. Further calculations including the year's keywords list, citation keywords list and root keywords list are all based on the document database queries.

From Fig. 8, it can be seen that paper 8 with a title of "Keyframe control of smoke simulation" from year 2003 is an important paper which is cited by other papers from year 2004 (paper 9 in yellow) to year 2014 (paper 7 in pink). If one path and one reference node are selected, the top rank terms will be displayed as shown in Fig. 11 where it can be found that these terms are highly relevant to the paper content.

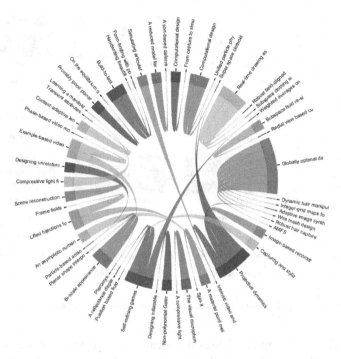

Fig. 9. Chord diagram visualization of citations of documents published in 2013.

6.3 Visualisation of Topic Trends

In the document analysis for citation tracing, topics of documents can be estimated by keywords. Keywords data linked with the publication year can be used to build topic trends. The topics were retrieved by the Latent Dirichlet Allocation (LDA)

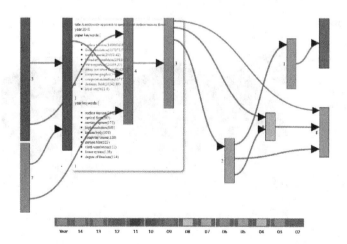

Fig. 10. Citation diagram with the maximum track depth of 8.

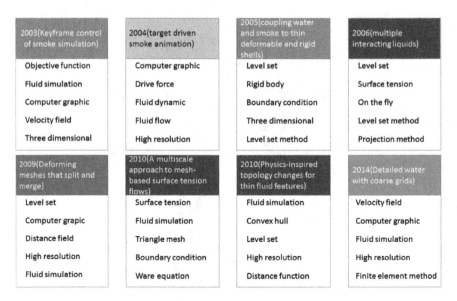

Fig. 11. Top 5 ranking keywords in each citation with the track length 8.

algorithm [39]. The algorithm assumes the documents were produced in a probabilistic generative model, which discovers the topics in every document or a corpus. Figure 12 shows a river-like trend visualisation [40] of six selected topics from year 2005 to 2014. The probabilities of topics are shown according to their proportion and represented by vertical length for each year. The six topics are represented by different colours. The visualisation is user interactive: mouse hovering on a topic at a certain year gives the topic name, year and probability value in that year in a tooltip.

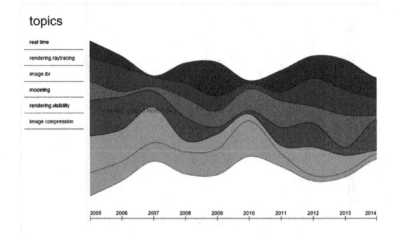

Fig. 12. Topic river visualization.

As far as ACM SIGGRAPH proceedings are concerned in this project, from the visualisation a decreasing number publication number can be seen in the topic of "image compression" while a chopping pattern shows in the topic of "real time". From these trends, the users can evaluate research topic trends.

7 Conclusions

In this paper, we present our work on text mining and data management on large scale scientific publications for collecting, tracking and presenting citations in a topic-enhanced way to facilitate scientific literature retrieval and research. Four data storages including a semantic repository, an index and search repository, a document repository and a graph repository are employed for efficient data management and fast information and graph retrieval. The keywords function of MAS API is used to collect keywords in 24 computer fields and to extract standard keywords frequency for each of the fields. Four levels of improved term weighting methods are designed to exploit term and topic characteristics in different aspects. The result citation network is stored in the graph database, accessed by efficient tailored queries and visualized in interactive chord diagrams and Sankey diagrams with enhanced topic information. The experiment results show that the combination of these techniques can efficiently store and index the publication data reliably to supply valuable information to support scientific research, which further helps researchers to derive meaningful insights of the published scientific resources more conveniently, enabling them to grasp technological change more quickly and hence assists new scientific discovery.

Acknowledgments. The research is supported by the FP7 Programme of the European Commission within projects Dr Inventor [FP7-ICT-611383] and CARRE [FP7-ICT-611140]. We would like to thank the European Commission for the funding and thank the project officers and reviewers for their indispensable support for both of the projects.

References

1. Wei, H., Zhao, Y., Liu, E., Wu, S., Deng, Z., Parvinzamir, F., Dong, F.: Management of scientific documents and visualization of citation relationships using weighted key scientific terms. In: DATA 2016, pp. 135–143 (2016)
2. Wei, H., Wu, S., Zhao, Y., Deng, Z., Ersotelos, N., Parvinzamir, F., Liu, B., Liu, E., Dong, F.: Data mining, management and visualization in large scientific corpuses. Edutainment **2016**, 371–379 (2016)
3. Grolinger, K., HigashinoEmail, W., Tiwari, A., Capretz, M.: Data management in cloud environments: NoSQL and NewSQL data stores. J. Cloud Comput. Adv. Syst. Appl. Adv. Syst. Appl. **2**(1), 2–22 (2013)
4. Kivikangas, P., Ishizuka, M.: Improving semantic queries by utilizing UNL ontology and a graph database. In: Proceedings of the 6th IEEE International Conference on Semantic Computing, pp. 83–86 (2012)
5. Neo4j. https://neo4j.com/

6. Tsai, F.S., Kwee, A.T.: Experiments in term weighting for novelty mining. Expert Syst. Appl. **38**(11), 14094–14101 (2011)
7. Debole, F., Sebastiani, F.: Supervised term weighting for automated text categorization. In: Proceedings of the 2003 ACM Symposium on Applied Computing, pp. 784–788. ACM Press (2003)
8. Zhang, Y., Tsai, F.S.: Combining named entities and tags for novel sentence detection. In: Proceedings of the WSDM Workshop on Exploiting Semantic Annotations in Information Retrieval (ESAIR 2009), pp. 30–34 (2009)
9. Manning, C.D., Raghavan, P., Schutze, H.: Introduction to Information Retrieval. Cambridge University Press, New York (2008)
10. Domeniconi, G., Moro, G., Pasolini, R., Sartori, C.: A study on term weighting for text categorization: a novel supervised variant of tf.idf. In: Proceedings of the 4th International Conference on Data Management Technologies and Applications, pp. 26–37 (2015)
11. Li, F., Pan, S.J., Jin, O., Yang, Q., Zhu, X.: Cross-domain co-extraction of sentiment and topic lexicons. In: Proceedings of the 50th Annual Meeting Association for Computational Linguistics: Long Papers (ACL 2012), vol. 1, pp. 410–419 (2012)
12. Domeniconi, G., Moro, G., Pasolini, R., Sartori, C.: Cross-domain text classification through iterative refining of target categories representations. In: Proceedings of the 6th International Conference on Knowledge Discovery & Information Retrieval (KDIR) (2014)
13. Alencar, A.B., Oliveira, M.C., Paulovich, F.V.: Seeing beyond reading: a survey on visual text analytics. Wiley Interdisc. Rev. Data Min. Knowl. Discov. **2**(6), 476–492 (2012)
14. Fu, S.: A survey on visual text analytics (2015). http://www.cse.ust.hk/~sfuaa/data/pqe.pdf
15. Federico, P., Heimerl, F., Koch, S., Miksch, S.: A survey on visual approaches for analyzing scientific literature and patents. TVCG (2016)
16. Zhao, D., Strotmann, A.: Analysis and Visualization of Citation Networks. Synthesis Lectures on Information Concepts Retrieval and Services, vol. 7(1) (2015)
17. Chen, C.: CiteSpace II: Detecting and visualizing emerging trends and transient patterns in scientific literature. J. Am. Soc. Inf. Sci. Technol. **57**(3), 359–377 (2006)
18. Zhang, J., Chen, C., Li, J.: Visualizing the intellectual structure with paper-reference matrices. IEEE TVCG **15**(6), 1153–1160 (2009)
19. Stasko, J., Choo, J., Han, Y., Hu, M., Pileggi, H., Sadana, R., Stolper, C.: Citevis: exploring conference paper citation data visually. Poster IEEE Vis. (2013)
20. Gorg, C., Liu, Z., Kihm, J., Choo, J., Park, H., Stasko, J.: Combining computational analyses and interactive visualization for document exploration and sense making in jigsaw. IEEE TVCG **19**(10), 1646–1663 (2013)
21. Doerk, M., Riche, N., Ramos, G., Dumais, S.: Pivotpaths: strolling through faceted information spaces. IEEE TVCG **18**(12), 2709–2718 (2012)
22. van Eck, N., Waltman, L.: CitNetExplorer: a new software tool for analyzing and visualizing citation network. J. Inf. **8**(4), 802–823 (2014)
23. Heimerl, F., Han, Q., Koch, S., Ertl, T.: CiteRivers: visual analytics of citation patterns. IEEE TVCG **22**(1), 190–199 (2016)
24. ACM SIGGRAPH. www.siggraph.org
25. MAS API. http://academic.research.microsoft.com/about/
26. Fensel, D., Hendler, J., Lieberman, H., Wahlster, W., Berners-Lee, T.: Sesame: An Architecture for Storing and Querying RDF Data and Schema Information. In: MIT Press eBook Chapters: Spinning the Semantic Web: Bringing the World Wide Web to Its Full Potential, pp. 197–222 (2005)

27. Cunningham, H., Maynard, D., Bontcheva, K., Tablan., V.: GATE: a framework and graphical development environment for robust NLP tools and applications. In: Proceedings of the 40th Anniversary Meeting of the Association for Computational Linguistics (ACL 2002), Philadelphia (2002)
28. Apach CouchDB. http://couchdb.apache.org/
29. Huang, H., Dong, Z.: Research on architecture and query performance based on distributed graph database Neo4j. In: Proceedings of the 3rd International Conference Consumer Electronics, Communications and Networks (CECNet), pp. 533–536 (2013)
30. Elasticsearch. https://www.elastic.co/products/elasticsearch
31. Elasticsearch attachment plugin. https://github.com/elastic/elasticsearch-mapper-attachments
32. pdfbox. https://pdfbox.apache.org/
33. CARRE. https://www.carre-project.eu/
34. ANNIE. https://gate.ac.uk/sale/tao/splitch6.html#chap:annie
35. Thakker, D., Sman, T., Lakin, P.: GATE Jape Grammar Tutorial, Version 1.0, A, Pictures, UK (2009)
36. Microsoft Academic Search (MAS) API. http://academic.research.microsoft.com/
37. D3. http://d3js.org/
38. Riehmann, P., Hanfler, M., Froehlich, B.: Interactive sankey diagrams. In: Proceedings of the IEEE Symposium on Information Visualization, pp. 233–240 (2005)
39. Blei, M., Ng, Y., Jordan, I.: Latent dirichlet allocation. J. Mach. Learn. Res. 3(4–5), 993–1022 (2003)
40. Havre, S., Hetzler, E., Whitney, P., Nowell, L.: Themeriver: visualizing thematic changes in large document collections. IEEE Trans. Vis. Comput. Graph. 8(1), 9–20 (2002)

Maturing Pay-as-you-go Data Quality Management: Towards Decision Support for Paying the Larger Bills

Jan van Dijk[1]([⊠]), Mortaza S. Bargh[1], Sunil Choenni[1,2],
and Marco Spruit[3]

[1] Research and Documentation Centre, Ministry of Security and Justice,
The Hague, The Netherlands
{j.j.van.dijk, m.shoe.bargh, r.choenni}@minvenj.nl
[2] Research Centre Creating 010, Rotterdam University of Technology,
Rotterdam, The Netherlands
r.choenni@hr.nl
[3] Department of Information and Computing Sciences, Utrecht University,
Utrecht, The Netherlands
m.r.spruit@uu.nl

Abstract. Data quality management is a great challenge in today's world due to increasing proliferation of abundant and heterogeneous datasets. All organizations that realize and maintain data intensive advanced applications should deal with data quality related problems on a daily basis. In these organization data quality related problems are registered in natural languages and subsequently the organizations rely on ad-hoc, non-systematic, and expensive solutions to categorize and resolve registered problems. In this contribution we present a formal description of an innovative data quality resolving architecture to semantically and dynamically map the descriptions of data quality related problems to data quality attributes. Through this mapping, we reduce complexity – as the dimensionality of data quality attributes is far smaller than that of the natural language space – and enable data analysts to directly use the methods and tools proposed in literature. Another challenge in data quality management is to choose appropriate solutions for addressing data quality problems due to lack of insight in the long-term or broader effects of candidate solutions. This difficulty becomes particularly prominent in flexible architectures where loosely linked data are integrated (e.g., data spaces or in open data settings). We present also a decision support framework for the solution choosing process to evaluate cost-benefit values of candidate solutions. The paper reports on a proof of concept tool of the proposed architecture and its evaluation.

Keywords: Data quality issues · Data quality management · Knowledge mapping · User generated inputs · Solution management

1 Introduction

Organizations and enterprises that realize and operationalize data intensive applications spend a lot of efforts and resources to deal with imperfections flaws, and problems in the (large and heterogeneous) datasets that they use as raw materials. For example, in our

© Springer International Publishing AG 2017
C. Francalanci and M. Helfert (Eds.): DATA 2016, CCIS 737, pp. 102–124, 2017.
DOI: 10.1007/978-3-319-62911-7_6

research center of the Dutch Ministry of Security and Justice, advanced applications are designed and deployed to produce insightful reports on judicial processes and crime trends for legislators, policymakers and the public. Example applications include Public Safety Mashups [1] and Elapsed Time Monitoring System of Criminal Cases [2]. These applications rely on various datasets – as collected and shared by our partner organizations – that are integrated by using data warehouse and data space architectures [3, 4]. Often such datasets contain inconsistent, imprecise, uncertain, missing, incomplete, … data values and attributes. Such problems in datasets may cause inaccurate and invalid data analysis outcomes, which can mislead data consumers eventually.

Upon detecting these problems in datasets, data analysts often report them in Issue Tracking Systems (ITSs) in order to address them later on categorically and collectively. There is no standard format for registering these problems and data analysts often describe them in natural languages in a quite freestyle form. For example, in a dedicated ITS, the data analysts in our organization have registered the following observed dataset problems: Not being able to process criminal datasets at a regional scale because the datasets were delivered at a national scale, not being able to carry out trend analysis due to lack of historical criminal data records, or not being able to run concurrent queries due to temporary datasets being distributed across various locations, a problem also reported in [5].

Because data analysts register observed dataset problems in natural languages, categorization of the registered problems based on their freestyle descriptions becomes tedious and challenging. On the one hand, problem descriptions belong to a "natural language space" of high dimensionality and complexity. On the other hand, finding some meaningful categories for these problem descriptions becomes another concern for data analysts. Having meaningful categories means that the problems in every category have similar solutions and can be resolved collectively. In practice, currently data analysts come up with ad-hoc, non-systematic, and expensive solutions to categorize and resolve registered problems.

Problems observed in datasets are generally related to Data Quality (DQ) issues. For instance, the problems in our datasets mentioned above are related to the DQ attributes of completeness and consistency. DQ is a field that is extensively studied in recent years, having a sound theoretical foundation and a rich set of solutions proposed in literature. It seems, therefore, promising to map the registered dataset problems to DQ issues. Hereby one can reduce complexity – as the DQ space dimensionality is far smaller than that of the natural language space – and make use of the DQ methods and tools proposed in literature directly. Mapping the registered problems to DQ issues, nevertheless, is not straightforward.

In this contribution, we aim at managing and resolving the dataset problems detected by data analysts through mapping them to DQ issues and making use of DQ management tools. (Note that we shall use terms "DQ related problems" and "DQ issues" to refer to dataset problems as described in natural language space and to refer to DQ issues as described in the DQ space, respectively.) To this end, we propose a functional architecture for

(a) Semantically mapping the linguistic descriptions of such problems to DQ issues,
(b) Automatically prioritizing the severity levels of DQ issues,

(c) Automatically categorizing DQ related problems according to the priority levels of the corresponding DQ issues, and

(d) Resolving DQ related problems based on their categories, which depend on the severity levels of the corresponding DQ issues.

When data analysts resolve these DQ related problems, they also carry out DQ management. As a by-product, therefore, the proposed architecture provides organizations with insight into their DQ issues in a dynamic (i.e., real-time) way, relying on user-generated inputs (i.e., the problem descriptions inserted by data analysts). From this perspective, our proposed architecture to map high-dimensional DQ related problems into low-dimensional DQ issues is inspired by [6] that aims "to bake specialized knowledge into the jobs of highly skilled workers" in order to take advantage of the rich body of knowledge in a field. By mapping the DQ related problems to DQ issues, we can look up the literature and tools that pertain to resolving the mapped DQ issues. Subsequently, the DQ related problems are solved according to the latest insights and tools. In [7–9] we presented a formal description and system architecture for an integrated system for resolving the problems observed in datasets based on DQ management principles. We evaluated the proposed architecture functionally and practically, the latter by design and realization of a proof-of-concept. The current work extends [9] with an additional framework for DQ solution management.

The paper starts with providing some background about DQ management and the related work in Sect. 2. Subsequently the motivations for and principles of our problem solving architecture are presented in Sect. 3 formally. The proposed architecture is validated by a proof-of-concept, as described in Sect. 4, where also some performance aspects are evaluated. Our conclusions are drawn and the future research is sketched in Sect. 5.

2 Background

This section gives some background information on the functional components of DQ management, outlines the motivations of the work, and provides an overview of the related work. For an overview of DQ management methodologies the interested reader is referred to [10].

2.1 Data Quality Management

DQ can be characterized by DQ attributes, which correspond to DQ issues in our notation mentioned above. DQ attributes are defined as those properties that are related to the state of DQ [11]. DQ Management (DQM) is concerned with a number of business processes that ensure the integrity of an organization's data during its collection, aggregation, application, warehousing and analysis [12]. As mentioned in [13]: "DQM is the management of people, processes, technology, and data within an enterprise, with the objective of improving the measures of Data Quality most important to the organization. The ultimate goal of DQM is not to improve Data Quality for the sake of having high-quality data, but to achieve the desired business outcomes that rely upon high-quality data." DQM can be decomposed into DQ assessment and DQ improvement functional components, as described below.

Flexible architectures and dynamic environments, e.g. data space architectures and open or linked data environments, are strongly user-oriented and are characterized by a pay-as-you-go data management approach [14]. Hence, DQ management in these situations is often performed from a local viewpoint. The "enterprise" as mentioned in [13] can in these architectures/environments be seen as a collaboration of organizations or data customers (e.g. data scientists), where the composition of the collaboration can vary, depending on the stakeholders of specific DQ related problems.

DQ Assessment. This component deals with determining which DQ attributes are relevant and the degree of their relevancy for an organization. As shown in Fig. 1 (i.e., the top half) DQ assessment encompasses identification, measurement, ranking, and categorization of the DQ attributes that are relevant for an organization's data, see [15, 16], where the latter reference provides a systematic approach to define DQ attributes. 'DQ attribute identification' is concerned with collecting possible DQ attributes from various sources like literature, data experts and data analysts. 'DQ measurement' and 'DQ attribute ranking' cover those processes that are for measuring and rating the importance of the identified attributes for the organization. 'DQ attribute categorization' deals with structuring the ranked attributes into a hierarchical representation so that the needs and requirements of the stakeholders like data managers, data experts, data analysts, and data consumers can be satisfied [15].

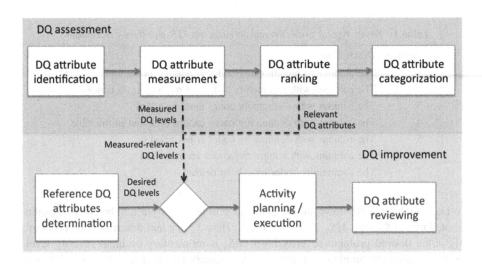

Fig. 1. Functional components of DQ management.

DQ Improvement. This component deals with continuously examining the data processing in an organization and enriching its DQ, given the relevant DQ attributes obtained from the DQ assessment. As shown Fig. 1 (i.e., the bottom half), the functional components of DQ improvement include 'reference DQ attribute determination', 'activity planning and execution', and 'DQ attribute reviewing' (partly adopted from [15]). 'Reference DQ attribute determination' identifies the organization's requirements related to the related DQ attributes, i.e., the desired DQ levels. 'Activity planning and execution' plans and carries out the activities required for improving the relevant DQ attributes to the desired level through, for example, executing a 'data cleansing' activity. Subsequently, one should also do 'DQ attribute reviewing' to validate these activities based on their dependency and measure the improved DQ attribute levels. The latter aspect of measurement can be seen as part of DQ assessment, see also [15].

2.2   Motivation

There are software products called Issue Tracking Systems (ITSs) to manage and maintain the lists of issues relevant for an organization; issues like software bugs, customer issues, and assets. Also in our organization, i.e., the Research and Documentation Centre (abbreviated as WODC in Dutch) of the Dutch Ministry of Security and Justice, we use such an ITS to keep track of the existing DQ related problems (Table 1). The WODC systematically collects, stores and enhances the Dutch judicial information directly or via its partner organizations [18]. Considering the diversity and distribution of our data sources, we often receive the corresponding datasets containing inconsistent, imprecise, uncertain, missing, incomplete, etc. data records and attributes. Our objective for registering DQ related problems is to keep track of how and whether (other) data analysts resolve these problems based on their severity and urgency.

Table 1. Seven typical problems registered in our ITS and their descriptions.

Problem	Description
1	The column with community codes is missing in the table
2	The columns with community codes are missing in all tables
3	The column with community codes must be added
4	The column with community codes cannot be found in the table
5	The column with community codes is not filled
6	The columns with community codes are not filled
7	The community codes have been deleted

Data analysts write down an encountered problem P_n by a number of parameters denoted by $P_n(X_n, DS_n, MS_n, PU_n); n : 1 \ldots N$. Here X_n is a text describing the problem, DS_n is the desired problem severity level, MS_n is momentary problem severity level, and PU_n represents problem urgency. The momentary problem severity level MS_n can be determined subjectively as perceived by the data analyst or objectively as measured based on some data specific parameters, by using for example the approach proposed in [19]. The data analyst determines the desired problem severity level DS_n subjectively. Both DS_n and MS_n are expressed in a real number between 0 and 1, where 1 means the problem severity is the highest. We assume that $0 \leq DS_n \leq MS_n \leq 1$ and the problem is resolved when $MS_n = DS_n$ or $MS_n = 0$, which in this case the problem can be removed from the ITS. Problems can have various impacts comparatively. Therefore the weigh factor PU_n – a real value between 0 and 1 where 1 means the highest urgency – is inserted by data analysis subjectively. Variable PU_n conveys the level of the problem's urgency compared with other reported problems. Let's denote the set of problems registered at the ITS by:

$$\{P_n(X_n, DS_n, MS_n, PU_n)|0 \le DS_n \le MS_n \le 1\} \text{ where } n : 1...N \qquad (1)$$

Figure 2 shows the functional components of a typical problem resolving system, status of which can be maintained in an ITS. Technical staffs - data quality managers or data analysts themselves, analyze the causes of a problem and its candidate solutions in order to choose a solution based on some trade-offs. Before, during and after the realization of a solution some Key Performance Indicators (KPIs) are used to measure the momentary problem severity levels so that the impact of devised solutions can be determined via the feedback loop. Although registered problems are related to DQ attributes, the textual definitions of problems are not specified in terms of DQ attributes due to lack of knowledge or interest about DQ concepts by data analysts.

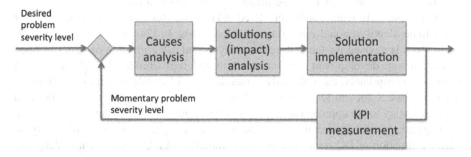

Fig. 2. A framework for resolving the DQ related problems registered at the ITS.

Furthermore, data analysts solve problems within certain boundaries (i.e. based on a certain priority or organizational limitations). Often, problems arise when data customers are working with the data. In these cases, a urgent solution is required for the dataset that is relevant for these data customers. Depending on how DQmanagement is organized, it is foreseeable that the budget for implementing the solution may come from the data customers that face the problem. This is especially the case in environments in which data management is not centrally organized, like data space architectures [3] and open data based applications/systems. We observe that, as a result of more user-oriented or pay-as-you-go data management, it becomes likely that problems are solved in a more local manner. Local KPIs ensure that the problem is solved in this particular context, but when the same problem occurs in a different context, it will be registered and handled as a new problem.

2.3   Related Work

As mentioned in Subsect. 2.2, ITSs are widely used for tracking and managing various issues relevant for an organization. The tracked issues range from software bugs in software development houses like Bugzilla [20] and JIRA [21], customer issues in customer support callcenters/helpdesks like H2desk [22], and assets in asset management companies like TOPdesk [23]. Software developers, customers, and employees of organizations use ITSs to report on the issues they face. These issues are reported in terms of the (detailed) description of the problem being experienced, urgency values (i.e., the overall importance of issues), who is experiencing the problem (e.g., external or internal customers), date of submission, attempted solutions or workarounds, a history of relevant changes, etc. Sometimes an issue report is called ticket due to being a running report on a particular problem, its status, and other relevant data with a unique reference number (as ITSs were originated as small cards within a traditional wall mounted work planning). Based on these reports, organizations take appropriate actions to resolve the corresponding problems. While there are many applications of ITSs for collaborative software development, including also management of announcements, documentation and project website, there are no applications of such systems for DQ management as we present in this contribution.

A possible feature that can be registered in ITSs is a user assigned label/tag in order to facilitate identifying and managing observed issues. In [24], for example, a visualization tool is devised for facilitating the analysis and categorization of issues in open source software development projects, based on such registered labels. Labelling, when it is done appropriately, can reduce the semantic space of registered issues and facilitate mapping these issues to DQ attributes. This means that labels and tags can be used complementary to our approach for an improved mapping of DQ problems to DQ issues.

DQ management approaches proposed in literature, on the other hand, often rely on offline estimation of DQ issues and/or offline inquiries of DQ requirements. Wang and Strong [15] propose a two-stage survey and a two-phase sorting method for identifying, ranking, and categorizing of DQ attributes in a given context. The authors developed a survey to produce a list of potential DQ attributes by a group of the participants of a workshop. Using another survey, the authors asked another group of the participants to rate the potential DQ attributes. In most organizations (including ours) gathering such a number of participants, i.e., data analysts, for surveying and sorting of DQ attributes is almost impossible due to being time consuming or having too few participants to produce valid results.

Woodall et al. [17] propose a so-called hybrid approach for DQ management. For a set of relevant DQ attributes, the approach assesses the required level of DQ improvement by comparing the current state to a reference state. The DQ management and improvement according to the hybrid approach remains very abstract because DQ diagnostics are based on some high level strategic concepts. Similarly to the hybrid approach, our DQ management is intertwined with operational level practices of data analysts who observe and resolve (DQ related) problems. Establishing this link in our proposal, however, delivers a pragmatically dynamic DQ management, which is not the case in the hybrid approach.

All researches related to DQ assessment depend on some DQ objectives, based on which a set of relevant DQ attributes are sought. For example, the Environmental Protection Agency (EPA) approach [25] relies on, among others, a review of DQ objectives, a preliminary review of potential anomalies in datasets, and a statistical method to draw quantitative DQ related conclusions from the data. Our study uses the idea of translating DQ problems into the DQ issues and objectives, but by considering 'all reported' problems in the datasets and not just a few reported anomalies as [25] does. Moreover, unlike [25] we don't rely on statistical methods exclusively and incorporate also the domain knowledge of data analysts. Pipino et al. [26] use the EPA methods and additionally incorporate a subjective DQ assessment. To this end, the authors use a questionnaire to measure the perceptions of the stakeholders (e.g., database administrators) on DQ attributes. Subsequently, the approach of [26] determines the root causes of data discrepancies and tries to improve DQ by solving these discrepancies. Also our proposal combines both subjective and objective perceptions of the stakeholders on DQ related problems, but we combine these perceptions at an operational level by using a problem solving system, and not on a DQ attribute or strategic level as [26] does. Eppler and Witting [27] use the EPA methods and adds some extra attributes to evaluate how pragmatic every DQ attribute can be realized. Unlike [27] we do not use any additional attribute to determine how pragmatic the DQ attributes are.

Possible ways of resolving data quality related problems are bound by several aspects. In the process of defining the data quality objectives, developed by EPA [25], every step describes some sort of boundary. Which solution method is applied depends on the organizational scope, budget, planning, etc. For instance, when a project experiences some problems with the timeliness of a specific dataset – i.e. the project needs up-to-date data – and no other project in the organization has the need or resources to invest in this problem, the problem might be fixed within project scope. From a strict data quality management perspective, in which high data quality is achieved when the data fit its intended use [15], the problem is solved. But in long term other projects might experience the same problem, and then it is inefficient to continue fixing the problem within project scope. Data quality managers must have the knowledge that this problem occurred before and at strategic level managers have to decide if this problem has to be dealt with in a more centralized way. This exceeds the scope of a single data quality problem or cluster of current problems, because it also involves those already solved problems. Lee et al. [28] mentioned in 2003 already that it is essential for improving data quality problem resolving to register how problems in the past were solved. However, to determine which problems are most urgent and which solutions are most appropriate, a lot of knowledge and expertise are required. The more this knowledge is put into operational use, the more the maturity of data quality management in an organization grows [29]. Mostly, this is done by implementing a data quality division or team [28] that can combine technical as well as organizational insights into solving data quality related problems. For instance, such a team shall remember those imperfect solutions that lead to recurring similar problems later on, so eventually a more sustainable solution can be sought. We extend our DQ management architecture in [9] with a knowledge-based framework that helps evaluating chosen solution methods and eventually assisting the choice of the (best) solution method for new problems. This way, also more flexible architectures like those for data spaces and open data communities can mature in their DQ management.

3 Proposed Approach

Figure 3 shows our proposed system architecture for resolving data quality related problems, which is described formally in [9]. We describe the key functional building blocks of this architecture, those marked with a *, in the following subsections.

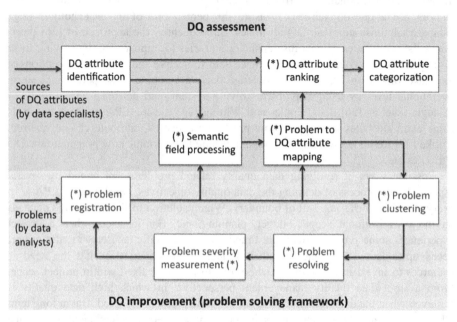

Fig. 3. Functional architecture of the proposed system for resolving DQ related problems based on DQ management.

In the last subsection we elaborate more on the solution choosing and propose a separate framework for solution management, mainly to support the Problem Clustering and Problem Resolving components. This framework allows us to guide the process of solution choosing, which in turn can lead to decision support on how to solve DQ related problems.

3.1 Data Quality Assessment

DQ assessment starts with a literature study by data specialists to enlist potential DQ attributes and ends up with categorizing the selected and ranked DQ attributes. The ranking of DQ attributes, which we innovatively base on the set of problems registered in the ITS, will be described in the following.

Semantic Field Processing. A semantic-field is a set of conceptually related terms [30]. Every semantic-field, which corresponds to only one DQ attribute in our setting, comprises a number of 'related terms'. Every related term, in turn, corresponds to a number of 'phrase sets'. Every phrase set comprises a number of phrases that appear in problem descriptions. The set of semantic-fields, related terms and phrase sets are summarized in a so-called 'Semantic-Field Processing Table (SFPT)' (Table 2). Formally, every DQ attribute DQ_m (where $m : 1...M$) can be described by a distinct semantic field S_m that consists of some semantic field attributes called related terms $RT_{m,i}$. In other words,

$$DQ_m \equiv S_m = \left\{ RT_{m,i} \middle| i : 1...M_m \right\} \tag{2}$$

where $m : 1...M$. In turn, every related term $RT_{m,i}$ can be described by some phrase sets $PS_{m,i,j}$ as

$$RT_{m,i} = \left\{ PS_{m,i,j} \middle| j : 1...M_{m,i} \right\} \tag{3}$$

where $m : 1...M; i : 1...M_m$. Every phrase set $PS_{m,i,j}$ comprises some set members/short phrases $PH_{m,i,j,k}$ as

$$PS_{m,i,j} = \left\{ PH_{m,i,j,k} \middle| k : 1...M_{m,i,j} \right\} \tag{4}$$

Domain experts define these semantic-fields, related terms, phrase sets, and short phrases in a way that the short phrases can be found in problem descriptions of data analysts; any related term can be related to only one semantic-field/DQ attribute; and any phrase set can be related to only one related term. Thus, as illustrated in Fig. 4, we assume that there is a tree structure among 'semantic fields', 'related terms', and 'phrase sets'. Due to the tree structure depicted above, there are no related terms that

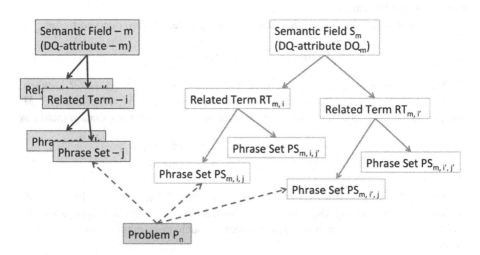

Fig. 4. An illustration of the hierarchical structure of semantic fields, related terms and phrase sets; and their relation to problems (the texts in grey blocks are intentionally abbreviated).

are common among semantic-fields/DQ attributes, and there are no phrase sets that are common among related terms.

$$
\begin{aligned}
RT_{m,i} &\neq RT_{m',i'} \qquad & \forall m \neq m' \text{ or } i \neq i' \\
PS_{m,i,j} &\neq PS_{m',i',j'} \qquad & \forall m \neq m' \text{ or } i \neq i' \text{ or } j \neq j'
\end{aligned}
\tag{5}
$$

Note that short phrases in phrase sets may appear in multiple phrase sets.

Table 2. Example of a semantic field-processing table (over the DQ attribute "completeness")

Phrase_1[a]	Phrase_2[a]	Related terms	DQ attribute (semantic field)
Is	Missed	Missing data	Completeness
Are	Missed	Missing data	Completeness
Be	Added	Adding data	Completeness
Is	Deleted	Lost data	Completeness
Are	Deleted	Lost data	Completeness

[a]*Derived from problem description. {Phrase_1, Phrase_2} is called a Phrase Set*

Problem to DQ Attribute Mapping. When a problem description contains all short phrases of a phrase set, one can map the problem to the corresponding related term and, in turn, to the corresponding DQ attribute uniquely. Based on Condition (5), phrase sets are unequal (see also the illustration in Fig. 4). This property and the hierarchical relation among phrase sets, related terms and semantic fields guarantee that every phrase set can identify only one related term, thus one semantic field/DQ attribute. As a problem description X_n may include more than one phrase sets, however, the corresponding problem P_n can be associated with more than one related term and thus to more than one DQ attribute.

Assume that the semantic fields identified for problem P_n are denoted by set

$$
S(P_n) \subseteq \{S_1, S_2, \cdots, S_M\}; n : 1 \ldots N
\tag{6}
$$

Then, problem P_n can be mapped to DQ attributes DQ_m if $S_m \in S(P_n)$, where $m : 1 \ldots M$. For problems P_n and DQ attributes DQ_m where $n : 1 \ldots N$ and $m : 1 \ldots M$, one can define the problem to DQ attribute mapping in terms of a association matrix as

$$
A = \left[a_{n,m} \right]_{N \times M} \text{where } a_{n,m} =
\begin{cases}
1 & \text{if } S_m \in S(P_n) \\
0 & \text{otherwise}
\end{cases}
\tag{7}
$$

Note that if $a_{n,m} = 0$ for all $m : 1 \ldots M$, i.e., when $S(P_n) = \emptyset$, then problems P_n cannot be mapped to any DQ attribute. In this case we say that the mapping for this problem has resulted in a miss. The number of such miss outputs should be zero ideally.

For improving DQ attributes, as we will see in the following sections, we need to take into account the momentary and desired severity levels of problems, i.e., the DS_n

and MS_n parameters of problem P_n registered in the ITS. Therefore, we define the *weighed association matrix* as

$$A_w = \left[aw_{n,m}\right]_{N \times M} \text{where } aw_{n,m} = a_{n,m} \cdot (MS_n - DS_n) \tag{8}$$

The problems registered in the ITS, furthermore, can have various urgency and importance levels, denoted by weight PU_n for problem P_n with a real value between 0 and 1 (remember that low or zero urgency issues are minor and should be resolved as time permits). Such a factor can be applied to Relation (8) by replacing $MS_n - DS_n$ with $PU_n.(MS_n - DS_n)$ to obtain the *extended weighed association matrix* as

$$A_{ew} = \left[aew_{n,m}\right]_{N \times M} \text{where } aew_{n,m} = a_{n,m} \cdot PU_n \cdot (MS_n - DS_n) \tag{9}$$

Note that the problems in the ITS are registered by data analysts, and therefore PU_n denotes the urgency of the problems from the viewpoint of the data analyst. This is a local worldview, because it is perceived from the viewpoint of a specific problem as observed by a specific data analyst in the field.

DQ Attribute Ranking. This functionality determines the priority values of DQ attributes based on the (extended weighted) association matrix, which is in turn derived from the problem descriptions, problem desired and actual severity levels, and/or problem urgencies. Given the (extended) weighted association matrix in Relation (8) or (9), the *dynamic DQ rank* of attribute DQ_m for $m : 1 \ldots M$ is defined as:

$$R_m^d = \frac{\sum_{n=1}^{N} aw_{n,m}}{\sum_{n=1}^{N} \sum_{m=1}^{M} aw_{n,m}} \text{ or } \frac{\sum_{n=1}^{N} aew_{n,m}}{\sum_{n=1}^{N} \sum_{m=1}^{M} aew_{n,m}} \tag{10}$$

As the elements of the (extended) weighted association matrix (i.e., $aw_{n,m}$ or $aew_{n,m}$) are dependent of the momentary problem severity level MS_n, which changes as problems are resolved by data analysts, the DQ rank in Relation (10) is a dynamic value depending on the problem resolving process. As a special case of DQ ranking in Relation (10), we define the *static DQ rank* based on the association matrix in Relation (7) for $m : 1 \ldots M$ by:

$$R_m^s = \frac{\sum_{n=1}^{N} a_{n,m}}{\sum_{n=1}^{N} \sum_{m=1}^{M} a_{n,m}} \tag{11}$$

The static DQ rank defined in Relation (11) is just dependent of having a problem in the ITS or not. The underlying assumption is that a problem is removed from the ITS as soon as it is resolved. This static DQ rank is called static because it does not change as the resolving of a problem progresses unless it is removed from the ITS.

3.2 Data Quality Improvement

Our DQ improvement largely corresponds to the problem-resolving system, as shown in Fig. 3. By solving the registered problems, data analysts also improve the corresponding DQ attributes and therefore carry out DQ management. DQ improvement comprises a number of functions, as shown in Fig. 3, which are elaborated upon in the following.

Problem Clustering. Registered problems can be clustered according to some criteria in order to reuse those solutions that address similar problems and, consequently, to yield efficiency and optimization. Our proposal for problem clustering is to use the associations among problems and DQ attributes because the resulting clusters can benefit from those DQ specific knowledge and solutions proposed in the literature. Data consistency problems for instance can be resolved by adopting a centralized architecture. Both data consistency and data completeness problems can be resolved by improving registration protocols or by implementing constraints at the physical database level (i.e. integrity and value-required ("not null") constraints for data consistency and data completeness problems, respectively).

As defined in Relations (7–9), the problem to DQ attribute mapping results in some (weighed) association values between pairs of (problem P_n, DQ attribute DQ_m) as follows:

$$(P_n, DQ_m) = \begin{cases} a_{n,m} & \text{see (7)} \\ aw_{n,m} = a_{n,m} \cdot (MS_n - DS_n) & \text{see (8)} \\ aew_{n,m} = a_{n,m} \cdot PU_n \cdot (MS_n - DS_n) & \text{see (9)} \end{cases} \quad (12)$$

We specify every problem P_n by the vector $((P_n, DQ_1), (P_n, DQ_2), \cdots, (P_n, DQ_M))$ in M dimensional DQ attribute space, where its elements are defined in Relation (12) for $m : 1 \ldots M$. We call these vectors as 'association vector', 'weighed association vector', or 'extended weighed association vector' of problem P_n, respectively.

The ((extended) weighed) association vectors are fed as inputs to the component 'problem clustering' as shown in Fig. 3. In order to find similarity between problems one can calculate the distance between every pair of such vectors, using for example the hamming distance or Euclidian distance. The pairwise distances can be used to cluster the corresponding problems. The resulting clusters encompass those problems that share similar behaviors in terms of DQ attributes. In order to address registered problems one can prioritize problem clusters, for example based on their sizes and weighs, and apply (and/or develop new) solutions that address these problem clusters according to the priority of the problem clusters.

Alternatively, one can *classify* problems in terms of existing solutions, instead of clustering them based on some behavioral similarity in the DQ attribute spaces. For example, assume a software tool resolves/addresses a specific subset of DQ attributes. Availability of such tools that are specific to a subset of DQ attributes inspires us to consider classifying the registered problems in terms of the DQ attributes that are addressed by some powerful software tools.

A solution may address multiple registered problems all together. When this occurs, applying the solution affects all corresponding MS_n and even the DS_n. In practice applying a solution may change the DS_n, which was initially inserted by a data analyst. For example, when implementing the solution it may turn out that the problem is (partial) infeasible to fix. In the following, we propose a method for choosing appropriate solutions, which resembles such a classification case.

Problem Resolving. Resolving of problems requires applying solutions, each of which encompasses a number of activities. Previously we specified problems in the DQ attribute space, i.e., by mapping problems to DQ attributes using the ((extended) weighed) association vectors and Relation (12). On the other hand, most solutions – including software tools and DQ improvement processes – can be characterized in terms of those DQ attribute issues that they address/resolve. Therefore, we propose to specify such solutions based on the DQ attributes that they address. To this end, assume every solution S_k is represented by a solution association vector $S_k = (s_{k,1}, \cdots, s_{k,m}, \cdots, s_{k,M})$ where for $m : 1 \ldots, M$ we have

$$s_{k,m} = \begin{cases} 1 & \text{if } S_k \text{ addresses } DQ \text{ attribute } DQ_m \\ 0 & \text{otherwise} \end{cases} \tag{13}$$

Here we assume solution S_k either addresses DQ attribute DQ_m or not, i.e., $s_{k,m}$ takes a binary value. One can alternatively assume a real value for parameter $s_{k,m}$ in interval $0 \leq s_{k,m} \leq 1$, denoting the fraction that solution S_k can (potentially) resolve the DQ attribute issue DQ_m in the organization. Hereto, for example, the approach of [19] can be used. Considering the dynamic or static rank of every DQ attribute, see Relations (10) and (11) respectively, one can define the normalized benefit of solution S_k for the organization as:

$$BF_k = \frac{1}{M} \begin{cases} S_k \cdot R^d = \sum_{m=1}^{M} s_{k,m} \cdot R_m^d & \text{if dynamic} \\ S_k \cdot R^s = \sum_{m=1}^{M} s_{k,m} \cdot R_m^s & \text{if static} \end{cases} \tag{14}$$

where upper scripts d and s demote dynamic and static DQ management, respectively.

On the other hand, one must balance the benefits of a solution, as characterized in Relation (14), against its costs. Various solutions inflict various costs on an organization. Let weight SC_k denote the normalized cost of solution S_k for the organization, by normalised we mean taking a real value between 0 and 1, where low or zero values represent those low or zero cost solutions. The cost-benefit value of a solution can be defined as

$$CB_k = SC_k - BF_k \text{ for } k : 1 \ldots K \tag{15}$$

Ideally one should prioritize solutions based on Relation (15) and apply those solutions that yield the lowest cost-benefit values as defined in Relation (15). We elaborate further on choosing the best solution in Subsect. 3.3.

Problem Severity Measurement. KPIs can be defined and used to measure the momentary severity of problems. As shown in Fig. 3, this functional block closes the loop of our current problem-resolving system and provides a feedback about the momentary status of registered problems, i.e., enables our dynamic DQ management.

In order to create objective KPIs we observe that often in practice DQ related problems are detected because some phenomena, for example the number of crimes committed per a time interval, are quantified differently from two (or more) data sources. Assume $X_t = \cdots, x_{t-1}, x_t, x_{t+1}, \cdots$ and $Y_t = \cdots, y_{t-1}, y_t, y_{t+1}, \cdots$ are time series that denote the measures of the same phenomenon using two different sources/datasets at consequent time intervals (yearly, monthly, daily etc.). Ideally, $x_t = y_t$ for all t, but due to DQ issues the data analyst observe discrepancies between these readings and reports the problem in the ITS. The difference time series $Z_t = X_t - Y_t = \cdots, x_{t-1} - y_{t-1}, x_t - y_t, x_{t+1} - y_{t+1}, \cdots$ can be a KPI in time intervals, as shown in Fig. 5. For our DQ management one can normalize the difference time series to derive problem severity level at a given moment t by

$$z_{t,\text{norm}} = \frac{|x_t - y_t|}{\max(x_t, y_t)}, \max(x_t, y_t) > 0 \tag{16}$$

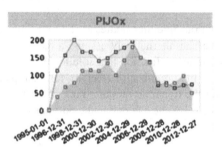

Fig. 5. Visualizations of two time series.

Sometimes it is more realistic to base problem severity level on the last l differences observed, i.e., on a history of measurements. Therefore, a smoothed problem severity level at a given moment t can be defined by

$$\bar{z}_{t,\text{norm}} = \frac{\sum_{i=t-l+1}^{t} |x_i - y_i|}{\sum_{i=t-l+1}^{t} (max(x_i, y_i) - t_h)} \tag{17}$$

where t_h is an appropriate threshold value – for example, it can be set as the possible minimum value for amount $max(x_i, y_i)$ over i (for example, when counting objects, this could be zero; for financial variables, the minimum could be negative).

The momentary or smoothed problem severity levels defined in Relations (16) and (17) can be visualized by a Gauge or Dial chart as shown in Fig. 6. Subjective measurements, where data analysts assign a problem severity level according to their insight at a given moment, can be another method for determining KPIs. Such a

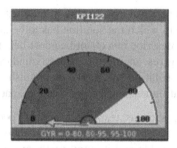

Fig. 6. Visualizations of the resulting-ratio dashboard.

subjective measurement can be useful when, for example, combining multiple and heterogeneous measures as defined in Relations (16) and (17).

3.3   A Framework for Data Quality Solutions Management

The data quality management architecture as described in [9] and in the previous subsections is sound for resolving DQ related problems. In this subsection we will elaborate more on the solution choosing, and propose a separate framework for solution management, mainly to support the Problem Clustering and Problem Resolving components. This framework allows us to formalize the process of solution choosing, which in turn can lead to decision support on how to solve DQ related problems.First, we will discuss for every step in the DQ improvement layerof our DQ management architecturethe impact on solution choosing.

- During Problem Registration a specific problem is registered in the ITS. In this stage only the local scope and urgency is known. Maybe solutions are proposed, but this will be also be done from a local perspective. For example, when data contain impossible values (i.e. outside the domain) the user might suggest setting these values to "unknown" as a temporary (local) solution.
- During Problem Clustering DQ analystsanalyze registered problems in order to cluster similar problems or classify them in terms of existing solutions. Due to the relation with DQ attributes (see Relation (13)), each problem is related to candidate solutions, i.e. the solution space. Problem Clustering defines the *theoretical* solution space, where the boundaries of the space are defined by the solutions that are applicable.
- During Problem Resolving, the boundaries of the solution space are narrowed down by the context of the problem. This is achieved by determining those solutions that are feasible given the current circumstances and by choosing the solution that has the best cost-benefit value.
- During Problem Severity Management some KPIs are developed to measure the momentary severity levels of resolved problems and compare them to the desired severity levels if possible. Note that these KPIs measure the effect of a chosen solution to a specific problem, and they do not take in account the possible effect of other solutions, nor the effect of the solution to other (not measured) problems.

Determining the cost-benefit valueof a solution requires a lot of knowledge about the domain and the context in which the solution has to be applied. This knowledge is often implicit and is hard to describe in a quantified cost-benefit value. In this paper, we propose an approach that gives an insight in the boundaries that influence chosen solutions, i.e. constrain the solution space. We distinguish the following boundaries, each of which cover several dimensions of the solution space:

1. Operational boundaries, such as resources (budget, people, software) and time frame. For instance, when you have a problem that should be solved by an expert, but there is no expert available and it is too costly or it takes too long to hire one, then this restriction forces to choose another – less optimal – solution.
2. Strategical boundaries, such as long-term business priorities. For instance, an organization has a certain budget for information management. It is a strategical choice how this budget is spent, e.g. focusing more on documentation or improving the data itself.
3. Organizational boundaries. An organization always has a certain role or scope regarding the data that are processed. For instance, an organization that is not involved in the registration of the data (e.g. a external research organization) will not be able to perform solutions that improve the registration process.
4. Domain-specific boundaries. Sometimes the domain in which data quality problems occur invoke limitations on the possibilities to improve data quality. For instance, in the criminal justice domain comparing police statistics to prosecution statistics (which is the next phase in the criminal justice chain) is challenging because exact matching of datasets is impossible due to a lack of common keys among datasets. This excludes several solutions for improving data quality at a record level.

Data analysts working in a specific domain and for a specific organization will have a good feeling for these boundaries when exploring the solution space for candidate solutions. However, a lot of these boundaries are flexible and should also be evaluated in a cost-benefit analysis. When the organization changes its strategy on information management, this can have immediate effect on the solutions. More interestingly, when the management of the organization has enough insight in the consequences of their strategical choices, they might change their strategy.

Data quality problems and their solutions, which are determined by the boundaries of the current solution space, can be seen as a solution model. Different solution models can be obtained by changing the boundaries of the solution space. The solution model with the best cost-benefit value might differ from the current one. In this way, i.e. by choosing an appropriate solution model, changing the DQ improvement strategy becomes part of the general information management strategy, which eventually leads to a more mature organization.

4 Proof of Concept

In this section we describe a proof of concept prototype for the proposed DQ management that is realized in our organization. Moreover, we shall elaborate on performance evaluation of its problem to DQ attribute mapping.

4.1   Implementation

Our realization of the proposed architecture includes problem registration, semantic field processing, problem to DQ attribute mapping, DQ attribute ranking, problem clustering, problem resolving, and problem severity measuring.

We used the Team Development environment of Oracle APEX as our ITS to enable data analysts to register the arising DQ related problems. The data log is stored in an Oracle DBMS (Database Management System). Currently, there are 334 problems registered together with their desired and momentary problem severity levels.

In order to determine the 'semantic-field processing table' for the registered problems, we use a heuristic as described below. Given a DQ attribute, the current implementation carries out two steps of (a) determining a list of the related terms for the semantic-field corresponding to the DQ attribute, and (b) syntactical decomposing of every related term to some phrases of smaller sizes that appear in problem descriptions. We assume that every phrase set $PS_{m,i,j}$ comprises at most two short phrases, i.e., $M_{m,i,j} \leq 2$ in Relation (4). Therefore, we shall sometimes use the term 'phrase pair' instead of 'phrase set'.

Assume that we have some potential DQ attributes derived from literature and that we have the actual problems descriptions registered in the ITS. In the first step of the heuristic we analyze every pair of (problem description, potential DQ attribute). When a problem description is conceptually related to a DQ attribute, then the conceptual formulation of the problem description is recorded as a related term. This related term has a smaller size than the corresponding problem description size. Iteration of this step results in two columns of the 'related terms' and the corresponding 'DQ attributes' in a semantic-field processing table. Lines (5) and (7) in the pseudo code below refer to this process. In the second step, every related term is decomposed into sets of smaller phrases that syntactically appear in problem descriptions. This results in another column 'phrase pair' in the semantic-field processing table. Lines (6) and (7) in the pseudo code below refer to this process.

```
▷ SFP is set of rows of the semantic-field processing ta-
ble
▷ rt is a related term
(1) SFP ← ∅
(2) for each problem description x do
(3)     for each potential DQ attribute dq do
(4)     if x refers to dq then
(5)     define rt as a conceptual formulation of dq
(6)     decompose x into (p1, p2)
(7)      if (p1, p2, rt, dq) ∉ SFP then
      SFP ← SFP ∪ (p1, p2, rt, dq)
```

Note that here some problems cannot be readily mapped to a DQ attribute. Moreover, the related terms obtained from the first stage are natural language terms. The syntactical decomposition of such natural language terms into phrase pairs can have more than one parsing tree [31]. For example, related term 'missing data' can be decomposed to phrase pairs {Is, Missed}, {Are, Missed} or {Are, Missing}.

Due to a prototype character of the current implementation, the clustering of problems and resolving problems according to their impacts/costs are currently based on a manual process. The measuring of the momentary severity level of problems is based on the described KPIs. The KPIs of complementary measurements, as defined in Relations (16) and (17), are defined in SQL terms and visualized by a dynamic PHP website. Currently, the ITS is deployed in another server and it is loosely coupled to the other components (as problem logs are downloaded as files). This slows down the communication between these two systems. In the future we intend to mitigate the communication speed of the current implementation.

4.2 Evaluation

Generic DQ management functionalities, which are identified in [17], are also represented in the proposed DQ management in this contribution. The proof of concept system has been realized, deployed, and used in our organization since early 2014. All functionalities of the realized system work as described in this contribution.

For performance evaluation here we report on the performance of our heuristic for the problem to DQ attribute mapping as the key system component in our problem solving system. Our heuristic cannot target all problems in the ITS because we start with DQ attributes and look at the problem descriptions in the ITS to identify the semantic-field of every DQ attribute (i.e., the related terms). Based on related terms our proof of concept seeks out the phrase pairs in a problem statement. As a result, this process may overlook some problems if for them no related term can be identified, thus failing to map such problems to DQ attributes. This overlooking could be due to not exhaustively searching the space of registered problems and DQ attributes or not describing problems expressively. Our search of related terms and phrases stops at a certain point due to practical reasons, for example, after finding a certain number of phrase-pairs.

Those problems that are (not) mapped to DQ attributes are called (un)targeted problems. In order to reduce the number of untargeted problems we iterated the heuristic described above to come up with the (new) related terms corresponding to some (potential) DQ attributes. These iterations reduced the number of untargeted problems sharply, as shown in Fig. 7. After a certain number of iterations, however, the number of untargeted problems did not decrease much. We suspect this is because the descriptions of the remaining problems are poorly written, which makes it difficult to associate them with any related term based on the syntax of these problem descriptions.

4.3 Discussion and Limitations

In this contribution we proposed to measure the severity level of the reported problems and map them to the corresponding DQ attribute levels. A way to measure the severity level of registered problems is to measure KPIs, which faces some challenges like defining effective, valid, and standardized performance indicators. For instance, a KPI based on measuring the hamming distance of 2 words can be ineffective. For instance, the words "Netherlands" and "Holland" are semantically closer than their Hamming

number of untargeted problems

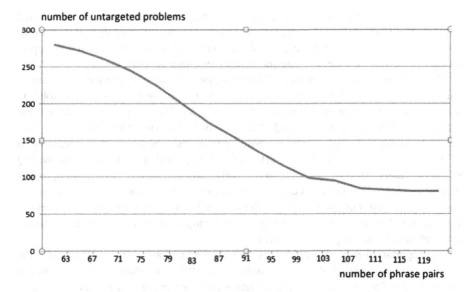

Fig. 7. Number of untargeted problems (vertical) in terms of the number of related terms.

distances when considering the cultural background of both words. Measuring semantic distances, on the other hand, is more challenging than measuring hamming distances.

An underlying assumption in our proposal is that data analysts of an organization register encountered problems in an ITS. In practice, users are not eager to register problems effectively and expressively. Organizations should encourage and train their employees to fill in such logging system so that the benefits of the proposed system can be harvested. Using tags and labels to mark DQ problems, [24] can further be explored to this end.

We proposed a data quality management approach to utilize user-generated inputs about DQ problems to carry out DQ management. For each functional component, furthermore, we proposed some simple (and heuristic) methods to realize the component's functionality. Due to modular property of the proposed DQ management approach, one can replace these methods by defining customized methods suitable for own organization and problem domain.

5   Conclusion and Further Research

In this contribution we presented the formal description and the system architecture of an integrated system for resolving the problems observed in datasets based on DQ management. The proposed architecture, moreover, results in a dynamic DQ management system, which relies on user generated data (i.e., data users/analysts who describe the DQ related problems they encounter in their daily practice). By managing DQ related problems encountered in an organization at an operational level, our proposal manages also the organization's DQ issues (i.e., realizes DQ management).

To this end, we semantically and dynamically map the descriptions of DQ related problems to DQ attributes. The mapping provides a quantitative and dynamic means to determine the relevant DQ attributes and the level of their relevancy, given the operational setting (i.e., the desired and momentary problem severity levels).

The realization of the proposed DQ management in our organization has given us insightful feedback on its advantages and limitations. As we envisioned, the solution bridged successfully the gap between the operational level (e.g., data analysts) and strategic level (e.g., managers) DQ stakeholders within our organization. To fully benefit from the potentials of the proposed architecture, however, it is necessary to encourage the users of datasets (i.e., data analysts) to provide their inputs about the DQ related problems that they encounter proactively and expressively. Through improving the problem registration process one can reduce the number of untargeted problems and guarantee their influence on dataset problem resolution and DQ management processes. It is for our future research to explore, for example, user awareness and training solutions, and to develop objective KPIs.

An important aspect of problem resolving in DQ management is to determine the capabilities and costs of candidate solutions. In this contribution we presented a framework for decision support in the solution choosing process by guiding the choice of the cost-benefit values of candidate solutions. In the future, we intend to formalize the proposed framework for DQ solution management and to develop a proof of concept in order to research how the framework can contribute to maturing DQ management in organizations with data intensive applications, and in collaboration with our system architecture for resolving DQ related problems.

Acknowledgements. Partial results of this work were presented earlier in [9]. Tables, figures and equations have their origin in this paper, unless stated otherwise.

References

1. Choenni, S., Leertouwer, E.: Public safety mashups to support policy makers. In: Andersen, K.N., Francesconi, E., Grönlund, Å., Engers, T.M. (eds.) EGOVIS 2010. LNCS, vol. 6267, pp. 234–248. Springer, Heidelberg (2010). doi:10.1007/978-3-642-15172-9_22
2. Netten, N., van den Braak, S., Choenni, S., Leertouwer, E.: Elapsed times in criminal justice systems. In: Proceedings of the 8th International Conference on Theory and Practice of Electronic Governance (ICEGOV), pp. 99–108. ACM (2014)
3. van Dijk, J., Kalidien, S., Choenni, S.: Smart monitoring of the criminal justice system. Government Information Quarterly. Elsevier (2016). doi:10.1016/j.giq.2015.11.005
4. Christoulakis, M., Spruit, M., van Dijk, J.: Data quality management in the public domain: a case study within the Dutch justice system. Int. J. Inf. Qual. 4(1), 1–17 (2015)
5. Birman, K.P.: Consistency in distributed systems. In: Guide to Reliable Distributed Systems, pp. 457–470. Springer, Heidelberg (2012)
6. Davenport, T.H., Glaser, J.: Just-in-time delivery comes to knowledge management. Harvard Bus. Rev. 80(7), 107–111 (2002)

7. Bargh, M.S., van Dijk, J., Choenni, S.: Dynamic data quality management using issue tracking systems. IADIS Int. J. Comput. Sci. Inf. Syst. **10**(2), 32–51 (2015). Isaias, P., Paprzycki, M. (eds.)

8. Bargh, M.S., Mbgong, F., Dijk, J. van, Choenni, S.: A framework for dynamic data quality management. In: Proceedings of the IADIS International Conference Information Systems Post-Implementation and Change Management, pp. 134–142 (2015)

9. Bargh, M., van Dijk, J., Choenni, S.: Management of data quality related problems - exploiting operational knowledge. In: Proceedings of the 5th International Conference on Data Management Technologies and Applications (DATA), pp. 31–42. SciTePress (2016)

10. Batini, C., Cappiello, C., Francalanci, C., Maurino, A.: Methodologies for data quality assessment and improvement. ACM Comput. Surv. **41**(3), 16–52 (2009). Article no. 16

11. Wand, Y., Wang, R.Y.: Anchoring data quality dimensions in ontological foundations. Commun. ACM **39**(11), 86–95 (1996). ACM

12. Davoudi, S., Dooling, J.A., Glondys, B., Jones, T.D., Kadlec, L., Overgaard, S.M., Ruben, K., Wendicke, A.: Data quality management model (2015 Update). J. AHIMA **86**(10), 62–65 (2015). expanded web version

13. Knowledgent: White Paper Series: Building a Successful Data Quality Management Program. http://knowledgent.com/whitepaper/building-successful-data-quality-management-program/. Accessed 31 Oct 2015

14. Halevy, A., Rajaraman, A., Ordille, J.: Data integration: the teenage years. In: Proceedings of the 32nd International Conference on Very Large Data Bases, pp. 9–16. VLDB Endowment (2006)

15. Wang, R.Y., Strong, D.M.: Beyond accuracy: what data quality means to data consumers. J. Manage. Inf. Syst. **12**(4), 5–33 (1996)

16. Price, R., Shanks, G.: A semiotic information quality framework. In: Proceedings of International Conference on Decision Support Systems (DSS), pp. 658–672 (2004)

17. Woodall, P., Borek, A., Parlikad, A.K.: Data quality assessment: the hybrid approach. Inf. Manage. **50**, 369–382 (2013)

18. Bargh, M.S., Choenni, S., Meijer, R.: Privacy and information sharing in a judicial setting: a wicked problem. In: Proceedings of the 16th Annual International Conference on Digital Government Research, pp. 97–106. ACM, New York (2015)

19. Jiang, L., Barone, D., Borgida, A., Mylopoulos, J.: Measuring and comparing effectiveness of data quality techniques. In: Eck, P., Gordijn, J., Wieringa, R. (eds.) CAiSE 2009. LNCS, vol. 5565, pp. 171–185. Springer, Heidelberg (2009). doi:10.1007/978-3-642-02144-2_17

20. Bugzilla Website. https://www.bugzilla.org. Accessed 31 Oct 2015

21. JIRA Software Website. https://www.atlassian.com/software/jira. Accessed 31 Oct 2015

22. H2desk Website, https://www.h2desk.com. Accessed 31 Oct 2015

23. TOPdesk Website. http://www.topdesk.nl. Accessed 31 Oct 2015

24. Canovas Izquierdo, J.L., Cosentino, V., Rolandi, B., Bergel, A., Cabot, J.: GiLA: GitHub label analyzer. In: IEEE 22nd International Conference on Software Analysis, Evolution and Reengineering, pp. 479–483, Montreal, Canada (2015)

25. Environmental protection agency: data quality assessment: a reviewer's guide, Technical report EPA/240/B-06/002, EPA QA/G-9R (2006)

26. Pipino, L.L., Lee, Y.W., Wang, R.Y.: Data quality assessment. Commun. ACM **45**(4), 211–218 (2012). ACM

27. Eppler, M.J., Wittig, D.: Conceptualizing information quality: a review of information quality frameworks from the last ten years. In: Proceedings of the Conference on Info Quality, pp. 83–96 (2000)

28. Lee, Y.: Crafting rules: context-reflective data quality problem solving. J. Manage. Inf. Syst. **20**(3), 93–119 (2003)

29. Ryu, K.S., Park, J.S., Park, J.H.: A data quality management maturity model. ETRI J. **28**(2), 191–204 (2006)
30. Kornai, A.: The algebra of lexical semantics. In: Mathematics of Language, pp. 174–199. Springer, Heidelberg (2010)
31. Mooney, R.J.: Learning for semantic parsing. In: Gelbukh, A. (ed.) CICLing 2007. LNCS, vol. 4394, pp. 311–324. Springer, Heidelberg (2007). doi:10.1007/978-3-540-70939-8_28

Generic and Concurrent Computation
of Belief Combination Rules

Frédéric Dambreville$^{(\boxtimes)}$

DGA MI/Lab-STICC, UMR CNRS 6285, Ensta Bretagne,
2 rue François Verny, Brest, France
submit@fredericdambreville.com
http://www.fredericdambreville.com

Abstract. As a form of random set, belief functions come with specific semantic and combination rule able to perform the representation and the fusion of uncertain and imprecise informations. The development of new combination rules able to manage conflict between data now offers a variety of tools for robust combination of piece of data from a database. The computation of multiple combinations from many querying cases in a database make necessary the development of efficient approach for concurrent belief computation. The approach should be generic in order to handle a variety of fusion rules. We present a generic implementation based on a map-reduce paradigm. An enhancement of this implementation is then proposed by means of a Markovian decomposition of the rule definition. At last, comparative results are presented for these implementations within the frameworks Apache Spark and Apache Flink.

Keywords: Map-reduce · Distributed data processing · Belief functions · Combination rules · Statistics

1 Introduction

Our hypothesis in this paper, in the perspective of a planned application, is that we have to request a database of partial information, which are registered and evaluated with a likelihood. An important point here, is that one single piece of information will not be sufficiently rich in general, in order to meet the requested criterion. However, some combinations of pieces of information may answer the request, and we want to process these information in order to alert the requester about some or several interesting combinations. The following picture is a rough sketch of what such request may be. In this example, an answer is requested with a minimal precision bound by the ellipsoid in the event space and with a minimal likelihood of 0.5. Then, three pieces of information are found which meet such requirement.

© Springer International Publishing AG 2017
C. Francalanci and M. Helfert (Eds.): DATA 2016, CCIS 737, pp. 125–140, 2017.
DOI: 10.1007/978-3-319-62911-7_7

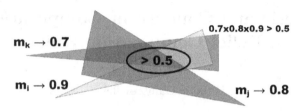

In real requests, the belief structure is much more complex than single propositions associated with likelihood. And while it is purposed that the sources of information are ideally independent, this requirement is generally only approximated.

Typical examples of such databases may be used in various Intelligence processes, where selected and filtered informations are collected and sourced. In general, such applications offer a rather good level of control, and the valuation of the information, in regards to the likelihood or to the sources independence, is much reliable. Other applicative contexts are provided by networked community of agents (*e.g.* especially opinion leaders in a social networks) or sensored sources (*e.g.* surveillance camera). But on such application cases, the control on the information sources is less reliable, and will need more robustness in the combination approach. In regards to some existing related works, we considered a formalism and combination processes based on belief functions. As a form of random set, belief functions come with specific semantic and combination rule able to perform the representation and the fusion of uncertain and imprecise informations. Moreover, the development of new combination rules able to manage conflict between data now offers a variety of tools for robust combination of piece of data from a database.

Belief Function Approaches from Related Works. Emerging works [1] have been done on the application of belief function to the analysis of interaction between agents of a social network on the basis of shared semantic content. These works are especially based [2] on evidential clustering of agents resulting in a fuzzy identification of communities. These clustering algorithms optimize the evidential similarities/dissimilarities between agents, but do not deeply involve combinations of beliefs and the semantic it could extract.

In the domain of surveillance, [3] have underlined eight challenge of a video network. Among them are the uncertainty of events, inconsistency or conflict between multiple sources, the composition of elemental events, and the scalability of the system. [4] have applied evidential networks to the problem of video surveillance in a controlled application (Smart transport). The structure implied by the networks makes possible an efficient reduction of the computational complexity.

Generic Rule Computation. In this paper, we assume that our information are preprocessed under the form of belief functions assigned to representative propositions. The produced information then takes the form of a large collection of belief functions, which are representative of agent/sensors viewpoints on a common topic. The combination of belief functions of a cluster by means of dedicated combination rules infers a refined analysis of the relative viewpoints, including agreement and disagreement on concomitant topics. Massive computation of such rules may be typically applied to the search of sets of complementary information in regards to an objective of fused answer with sufficient joint likelihood and precision requirement.

Beside, in order to deal with the information conflict and with relaxed hypotheses on the sources independence, the combination rules should be chosen amongst robust combinations and take into account the context and the structure of the information. The issue of the generated conflict has been challenged by many evolutions of the historical conjunctive and Dempster-Shafer rules [5–9], and the domain now offers a variety of solutions. For these reason our implementations should be generic in order to allow the use of dedicated combinations rules, and should be concurrent for addressing a large collection of cases.

The development of generic implementation [10] of combination rules is a challenge in itself. Our work consider both issues by extending a previous work [10] dedicated to the generic implementation of rules. In the continuation of this work and of our recent contribution, [11], this paper considers the problem of parallel and pooling computation by factoring the combination process, and a Map-Reduce [12] approach is proposed within the two computation framework: Apache Spark [13] and Apache Flink [14]. In order to factor some combination rules, new algebraic structures (e.g. multisets in the case of Dubois & Prade rule) are used as a processing space instead of the propositional framework of belief functions.

Section 2 introduces basic concept on belief functions. Section 3 presents our new contribution for a parallel and pooling implementation of combination rules; previous work [10] is also introduced. Section 4 presents some limited tests.

2   Belief Functions

As a form of random set, belief functions come with specific semantic and combination rule able to perform the representation and the fusion of uncertain and imprecise informations. Belief functions are representations of imprecise and uncertain information over an algebraic framework, a lattice in its most general form. Most authors consider belief functions over powersets, as a Boolean algebra, and this is our hypothesis here. From now is given the finite set Ω, the universe or frame of discernment.

2.1   Belief Assignments

The imprecise and uncertain information are characterized by basic belief assignment (bba), m, over propositions of the framework. Thus, a bba is defined as the attribution of pieces of belief to subsets of Ω:

$$m \geq 0 \text{ and } \sum_{X \subseteq \Omega} m(X) = 1. \tag{1}$$

In case of *closed world hypothesis*, it is assumed that the belief put on empty set is zeroed, *i.e.* $m(\emptyset) = 0$. This paper does not discuss further about such hypothesis, but its involvement does not imply a difficult generalization. However, we refer subsequently to $m(\emptyset)$ as the conflict related to m, and consider rules which redistribute the conflict.

Given M sources providing information by means of bba $m_{1:M}$, the fusion of these information are computed by combinations rules. In the case of a closed world, the reference combination rule of Dempster-Shafer [15, 16] is derived from the conjunctive rule by means of a normalization based on the conflict. Without lost of generality [17], Dempster-Shafer rules could be rewritten as a conjunctive rule without normalization in the case of open world.

2.2   Combination Rules

Given two bbas, m_1, m_2, from two sources of information, the conjunctive combination of m_1 and m_2 is defined by:

$$m_1 \oslash m_2(X) = \sum_{Y_1, Y_2 : Y_1 \cap Y_2 = X} m_1(Y_1) m_2(Y_2). \qquad (2)$$

This rule only works for open world hypothesis, since it is possible to have $m_1 \oslash m_2(\emptyset) > 0$ while $m_1(\emptyset) = m_2(\emptyset) = 0$.

By interpreting $m_1 \oslash m_2(\emptyset)$ as a measure of conflict and redistributing it, many alternative rules have been proposed. The development of new combination rules able to manage conflict between data now offers a variety of tools for robust combination. For example, Dubois & Prade rule [5], PCR5 rule and PCR6 rule [7, 9] implement different redistribution schemes:

- **Dubois & Prade Rule.** This rule extends the conjunctive rule by redistributing disjunctively the conflict. Appendix A.1 presents its original definition,
- **PCR Rules.** The PCR combination rules, pioneered by Dezert and Smarandache [7], are based on a local proportional redistribution of the conflict. Appendix A.2 presents the original definition of PCR6 rule proposed by Martin and Osswald.

As an example of concurrent implementation, we consider Dubois & Prade rule, generalized to many sources, but the approach is generic and addresses a potentially large scope of rules. Although we do not investigate implementations for PCR rules here, we present some theoretical elements about PCR6 for such a concurrent implementation.

3   Implementations

The bba $m_{1:M}$, to be fused, are taken amongst a set of bba $\tilde{m}_{1:N}$, where $2 \le M \ll N$. Typically, $m_i = \tilde{m}_{j[i]}$, where the *selection* map $j \in \{1 : N\}^{\{1:M\}}$ is injective in general. The combination of $m_{1:M}$ is done for selection j. The number of concerned selections could be very large. Our main concern and challenge is to implement the computation of combination rules for all selections as a distributed process. A *Map-Reduce* approach [12] is considered for this computation. A first approach is based on a previous work.

Section 3.1 introduces the generic formalism of referee functions [10] for defining combination rules. On this basis, Sect. 3.2 presents a Map-Reduce implementation of the combination rules. Section 3.3 enhances the formalism of referee functions with Markov properties, and improves the definition of the combination rules, with recursive computational properties. On this basis, Sect. 3.4 presents a Map-Reduce and recursive implementation.

Notation: Indicator functions are defined by:

$$I[P] = \begin{cases} 0 \text{ if } P \text{ is } \textbf{false}, \\ 1 \text{ if } P \text{ is } \textbf{true}. \end{cases} \tag{3}$$

3.1   Formulation Based on Indicators

[10] introduced a generic formulation of combination rules by means of conditional functions (*referee functions*), which have a computational meaning as indicator functions.

Common Principle. Combinations of bba $m_{1:M}$ is expressed under the form:

$$\oplus[m_{1:M}|F](X) = \sum_{Y_{1:M} \subseteq \Omega} F(X|Y_{1:M}; m_{1:M}) \prod_{i=1}^{M} m_i(Y_i). \tag{4}$$

In this formulation, a distinction is made between the rule processing expressed by the summation, and the rule definition which is expressed by the conditional indicator F. It is easy, then, to imply a generic distributed implementation of this summation, and we propose an implementation within both frameworks, Apache Spark [13] and Apache Flink [14]. This generic definition by means of indicator function is quite general however, as shown in [10], and typically, there are referee functions defined for conjunctive or disjunctive rules, D&P rules, PCR6 rule, and more. For the concern of this paper, we present the referee functions related to D&P rules and to PCR6, but only implement D&P.

Alternative Definition of D&P Rrule. The rule of Dubois and Prade [5] is defined by redistributing the conflict on the disjunction of the best consensuses:

$$m_1 \oplus_{DP} \cdots \oplus_{DP} m_M = \oplus[m_{1:M}|F_{DP}], \tag{5}$$

where:

$$F_{DP}(X|Y_{1:M}; m_{1:M}) = I\left[X = \arg\max_{\omega \in \Omega} \sum_{i=1}^{M} I[\omega \in Y_i]\right]. \tag{6}$$

Subset $\arg\max_{\omega \in \Omega} \sum_{i=1}^{M} I[\omega \in Y_i]$ matches the best vote derived from the belonging to propositions $Y_{1:M}$.

Alternative Definition of PCR6 Rule. PCR6 rule, in its general form, will test the full consensus, *i.e.* whether $\bigcap_{i=1}^{n} Y_i \neq \emptyset$ or not. If the full consensus works, it is returned, otherwise PCR6 will chose randomly amongst the hypotheses $Y_{1:n}$ in proportion to their individual bba.

$$m_1 \oplus_{PCR6} \cdots \oplus_{PCR6} m_M = \oplus \left[m_{1:M} \,\middle|\, F_{PCR6} \right], \tag{7}$$

where:

$$F_{PCR6}(X|Y_{1:M}; m_{1:M}) =$$

$$I \left[\bigcap_{i=1:M} Y_i \neq \emptyset \right] I \left[X = \bigcap_{i=1:M} Y_i \right]$$

$$+ I \left[\bigcap_{i=1:M} Y_i = \emptyset \right] \frac{\sum\limits_{i=1:M} m_i(Y_i) I \left[X = Y_i \right]}{\sum\limits_{i=1:M} m_i(Y_i)} \tag{8}$$

Computational Issues. There are actually two aspects to be considered, since the computation may be computing-intensive as well as data-intensive. On the one hand, it may be computing intensive, since a belief assignment is a vector of dimension $2^{\mathrm{card}(\Omega)}$; without approximation, the complexity of any belief computation increases dramatically with the size of the frame of discernment, and this issue is worsened with the number of bba to be combined. As a perspective of a distributed intensive computation of the rules, is the possibility to handle complex belief representations and their combinations for specific applicative use or conceptual studies. On the other hand, the computation may be data intensive, in the case of multiple combinations among a large collection of bba, typically issuing from local processing of a collection of sources of information. In this kind of application, the many sources of information produce pieces of data, from which knowledges are extracted by the local process in the form of bba in the context of a given logical frame. Then, the extracted bba are combined according to a combination plan, characterized by a selection function, in order to evaluate the compatibility of the sources or evaluate a confirmed knowledge. The combination plan generally implies a large amount of combination cases. Although our approach may be applied to both case, our preliminary tests focus on the second scenario. Now, we do not address here the question of the extraction, but only the question of the combination.

3.2   Map-Reduce Implementation

We implemented the generic fusion process (4) by means of a Map-reduce principle [12]. This implementation has been made respectively by means of frameworks Apache Spark [13] and Apache Flink [14].

Outline of the Implementation. We present the implementation on the basis of Spark formalism, where the computation flow is formalized by means of Resilient Distributed Dataset (RDD). The small difference with Flink implementation is immaterial in terms of scala commands. The computation follows two steps.

Mapping Steps. In these steps, the inner computations are done, that is the joint belief assignments, $\prod_{i=1}^{M} m_i(Y_i)$, and the definition maps, $F(X|Y_{1:M}; m_{1:M})$. These computations are derived for all considered selection maps j and all possible non-zero propositional combinations, $Y_{1:M}$. The amount of data is potentially exponential with M.

- Define the set of *selection* maps $j \in \{1 : N\}^{\{1:M\}}$ to be computed as a RDD of list, that is J: `RDD[List[Int]]`. For this purpose, method `flatMap` is applied to an iterator describing j,
- From selection map j and the definition of bba $\tilde{m}_{1:N}$, map to the collection of tuples:

$$\left(j, \left(Y_{j(i)}, \tilde{m}_{j(i)}\right)_{i=1:M}\right).$$

Only cases, with non zero values for $\tilde{m}_{j(i)}(Y_{j(i)})$, are considered. Methods `join` and `flatMap` are thoroughly used in this process, resulting in:

    M: RDD[(List[Int], List[(U,U=>Double)])],

where generic type U is used for encoding subsets,
- Referee function is applied through method `flatMap`, and yields the collection of tuples:

$$\left((j, X), F\left(X \middle| Y_{j(1:M)}, \tilde{m}_{j(1:M)}\right) \prod_{i=1:M} \tilde{m}_{j(i)}\left(Y_{j(i)}\right)\right),$$

as the RDD, FM: `RDD[((List[Int], U),Double)]`,

Reducing Step. In this step, all inner computations are summed up according to summation, $\sum_{Y_{1:M} \subseteq \Omega}$.

- At last RDD FM is reduced by key (j, X) with the addition operator. As a result, the combined bba are obtained as the collection of tuples:

$$(j, (X, \oplus [\tilde{m}_{1:M}| F](X))).$$

Method `reduceByKey` is used with +, then yielding:

    FusedM: RDD[(List[Int], (U,Double))].

At this time, the caching strategy is not monitored, and only RDD J, defining the selection, and RDD FM, defining the final result, are persistent.

Computational Issues. While this approach makes possible a fully distributed computation of the inner elements of the combination during the mapping steps, the amount of cases kept in memory increases exponentially with the number of sources to be combined. Even with a triple combination, the approach consumes a lot of memory, especially when the set of selected combinations is densely connected. In order to address this issue, we propose in next section to bring out and to implement a Markovian property of the rule definition.

3.3   Recursive Formulation

In general, the logical propositions do not contain sufficient information for a recursive definition of the rule. The main principle of our Markovian approach is to project the proposition on a computation space, which will indeed vehicle sufficient information for a recursive definition.

Common Principle. Sets Ψ and Λ are defined for intermediate computations. The rule is defined from three finite conditional functions:

$$(\lambda_i, Y_i, m_i) \mapsto \sigma(\lambda_i|Y_i; m_i), \tag{9}$$

$$(\psi_n, \lambda_{1:n}) \mapsto R(\psi_n|\lambda_{1:n}), \tag{10}$$

$$(X, \psi_M) \mapsto \pi(X|\psi_M), \tag{11}$$

for $X, Y_i \subseteq \Omega$, $\psi_n \in \Psi$, $\lambda_i \in \Lambda$ and $1 \leq i, n \leq M$. Functions σ and π are respectively forward and backward projectors from the space of proposition 2^{Ω} to the spaces of computation, Λ and Ψ. Function R is a *referee function within the spaces of computation*. In σ, parameter m_i is the bba related to source i, but any other contextual knowledge could be considered. Based on triplet $[\sigma, R, \pi]$, rule $\oplus[\sigma, R, \pi]$ is defined as a composition of conditional inferences:

$$\oplus\left[m_{1:M}\,\middle|\,\sigma, R, \pi\right](X) = \sum_{\psi_M \in \Psi} \pi(X|\psi_M)$$
$$\sum_{\substack{Y_{1:M} \subseteq \Omega \\ \lambda_{1:M} \in \Lambda}} R(\psi_M|\lambda_{1:M}) \prod_{i=1}^{M} \left(m_i(Y_i)\sigma(\lambda_i|Y_i; m_i)\right), \tag{12}$$

for all $X \subseteq \Omega$. Owing to definition (12), it is noticed that, although Ψ and Λ may be *infinite* sets, the summations are actually finite: the values are zeroed except for a finite number of them. On such definition, the main computation burden comes from the conditional inference $R(\psi_n|\lambda_{1:n})$, while other inferences are more or less easily factorized. In order to reduce the computational burden, a Markov hypothesis is made on R by introducing conditional function ρ:

$$R(\psi_{n+1}|\lambda_{1:n+1}) = \sum_{\psi_n \in \Psi} \rho(\psi_{n+1}|\psi_n, \lambda_{n+1}) R(\psi_n|\lambda_{1:n}), \tag{13}$$

for $\psi_{1:M} \in \Psi$, $\lambda_{1:M} \in \Lambda$ and $1 \leq n < M$. Under this hypothesis, $\oplus[m_{1:M}\,|\,\sigma, R, \pi]$ is computed recursively:

1. For $\lambda_{1:M} \in \Lambda$ and $i = 1 : M$, compute projection:

$$\mu_i(\lambda_i) = \sum_{Y_i \subseteq \Omega} m_i(Y_i)\sigma(\lambda_i|Y_i; m_i), \tag{14}$$

2. Compute $\oplus [m_{1:M}|R]$ recursively within space Ψ:

$$\oplus [m_1|R](\psi_1) = \sum_{\lambda_1 \in \Lambda} \mu_1(\lambda_1)R(\psi_1|\lambda_1), \tag{15}$$

$$\oplus [m_{1:n+1}|R](\psi_{n+1}) = \sum_{\lambda_{n+1} \in \Lambda} \mu_{n+1}(\lambda_{n+1})$$

$$\sum_{\psi_n \in \Psi} \rho(\psi_{n+1}|\psi_n, \lambda_{n+1}) \oplus [m_{1:n}|R](\psi_n), \tag{16}$$

for $\psi_{1:M} \in \Psi$, $\lambda_{1:M} \in \Lambda$ and $1 \le n < M$,
3. Compute backward projection for all $X \subseteq \Omega$:

$$\oplus [m_{1:M}|\sigma, R, \pi](X) =$$

$$\sum_{\psi_M \in \Psi} \pi(X|\psi_M) \oplus [m_{1:M}|\sigma, R](\psi_M). \tag{17}$$

Combined with *Map-Reduce*, the recursion improves the efficiency of the distributed implementation.

Recursive Definition of D&P Rule. *Unprojected* definition (6) is not directly compatible with a recursive decomposition. In order to compute the sources consensuses in a Markov decomposition, multisets, mapping from within Ω, are used as intermediate parameters.

$$m_1 \oplus_{DP} \cdots \oplus_{DP} m_M = \oplus [m_{1:M}|\sigma_{DP}, R_{DP}, \pi_{DP}], \tag{18}$$

where σ_{DP} is canonical map from sets to multisets, π_{DP} maps backward from multisets to *top* sets, and R_{DP} evaluates the vote by adding on the multisets:

- $\Lambda_{DP} = \Psi_{DP} = \mathbb{N}^\Omega$,
- $\sigma_{DP}(\lambda_i|Y_i; m_i) = I \left[\lambda_i = [I [\omega \in Y_i]]_{\omega \in \Omega} \right]$,
- $\pi_{DP}(X|\psi_M) = I \left[X = \arg \max_{\omega \in \Omega} \psi_M(\omega) \right]$,
- $R_{DP}(\psi_n|\lambda_{1:n}) = I \left[\psi_n = \sum_{i=1}^{n} \lambda_i \right]$.

Markov decomposition of R_{DP} comes easily:

$$\rho_{DP}(\psi_{n+1}|\psi_n, \lambda_{n+1}) = I [\psi_{n+1} = \psi_n + \lambda_{n+1}]. \tag{19}$$

Recursive Definition of PCR6 Rule. The definition by *unprojected* referee functions, (8), is not directly compatible with a Markov decomposition of the referee function. A Markov decomposition is however possible, but complex, by projecting to the space of computation, $[0, M] \times 2^{\Omega} \times 2^{\Omega}$, thus allowing the memorization of the normalization term $\sum_{j=1:M} m_j(Y_j)$ as well as the propagation of both the full consensus and the individual propositions. However, this decomposition needs a quantized approximation of interval $[0, M]$ in order to handle the computation. This complex process is still investigated.

3.4   Recursive Implementation

Recursive Map-Reduce implementations of the rules are similar to the non-recursive approach described in Sect. 3.2. But in this case, each recursive step is computed by means of a Map-Reduce sequence.

Projection Steps. Each bba of $(\widetilde{m}_k)_{k=1:N}$ is computed and mapped into its projection $(\widetilde{\mu}_k)_{k=1:N}$, by a map step and a reduce step:

– All bba are computed as a collection of tuples:

$$(k, (Y_k, \widetilde{m}_k(Y_k))),$$

as Bba: `RDD[(Int, (U,Double))]`. Generic type U is used for encoding subsets. For this purpose, `flatMap` is applied to an iterator of the bba,
– From Bba, the projected weights are then computed as a collection of tuples:

$$(k, (\lambda, \sigma(\lambda|Y_k; \widetilde{m}_k)\widetilde{m}_k(Y_k))).$$

RDD, `MapBba`: `RDD[(Int, (L,Double))]`, is obtained by applying `flatMap` to Bba. Generic type L is used for encoding Λ-parameters,
– `MapBba` is reduced by key (k, λ) with the addition operator. As a result, the projected bba, $\widetilde{\mu}_{1:N}$, are obtained as the collection of tuples:

$$(k, (\lambda, \widetilde{\mu}_k(\lambda))).$$

Method `reduceByKey` is used with +, yielding:
`ProjBba`: `RDD[(Int, (L,Double))]`,

Recursive Steps. First stage (15) is computed:

– For all prefixes, $j(1)$, of a selection map j, `ProjBba` is mapped to the collection of tuples:
$$(j(1), (\psi, \widetilde{\mu}_{j(1)}(\lambda)R(\psi|\lambda))).$$

`FusMapBba`: `RDD[(List[Int], (P,Double))]`, is obtained by applying `flatMap` to `ProjBba`. Generic type P is used for encoding Ψ-parameters,

- FusMapBba is reduced by key $(j(1), \psi)$ with the addition operator. As a result, values $\oplus \left[\widetilde{m}_{j(1)} \middle| R \right]$, are obtained as the collection of tuples:

$$\left(j(1), \left(\psi, \oplus \left[\widetilde{m}_{j(1)} \middle| R \right] (\psi) \right) \right).$$

Method `reduceByKey` is used with `+`, yielding:
    `FusProjBba: RDD[(List[Int], (P,Double))],`
- Set $n \leftarrow 1$,

and subsequent stages (16) are computed until $n = M$:

- Set $n \leftarrow n + 1$,
- For all prefixes, $j(1 : n)$, of a selection map j, FusProjBba is mapped to the collection of tuples:

$$\left(j(1 : n), \left(\psi', \widetilde{\mu}_{j(n)}(\lambda) \rho(\psi'|\psi, \lambda) \ \oplus \left[\widetilde{m}_{j(1:n-1)} \middle| R \right] (\psi) \right) \right).$$

FusMapBba: RDD[(List[Int], (P,Double))], is obtained by applying `flatMap` to ProjBba.
- FusMapBba is reduced by key $(j(1 : n), \psi)$ with the addition operator. Values $\oplus \left[\widetilde{m}_{j(1:n)} \middle| R \right]$, are obtained as the collection of tuples:

$$\left(j(1 : n), \left(\psi, \oplus \left[\widetilde{m}_{j(1:n)} \middle| R \right] (\psi) \right) \right).$$

Method `reduceByKey` is used with `+`, yielding:
    `FusProjBba: RDD[(List[Int], (P,Double))],`

Backward Projection Steps. At last, the combined bba are obtained from FusProjBba:

- For all selection maps j, FusProjBba is mapped to the collection of tuples:

$$\left(j, \left(X, \pi(X|\psi) \ \oplus \left[\widetilde{m}_{j(1:M)} \middle| \sigma, R \right] (\psi) \right) \right).$$

Then, FM: RDD[(List[Int], (U,Double))], is obtained by applying `flatMap` to FusProjBba,
- FM is reduced by key (j, X) with the addition operator. Combined bba are obtained as the collection of tuples:

$$\left(j, \left(X, \oplus \left[m_{1:M} \middle| \sigma, R, \pi \right] (X) \right) \right).$$

Method `reduceByKey` is used with `+`, then yielding:
    `FusedM: RDD[(List[Int], (U,Double))].`

4   Testing Cases

All tests have been made for the rule of Dubois & Prade.

4.1   Tests Presentation and Results

The tests have been done on a virtual machine processed on a bi-processor computer:

Memory	Processor	Frequency
23 Gio	Xeon X5690	3.67 GHz

The virtual machine is defined with 20 threads and 20 Gio. The tests have been processed with different threading:

Threads	Memory
1, 2, 4, 8, 12, 16	18 Gio

A collection of N bba is first generated randomly on set $\Omega = \{a, b, c, d\}$. The sampling method, based on uniform particles generation over $2^{\Omega} \setminus \{\emptyset\}$, tends to favor positive bba over $2^{\Omega} \setminus \{\emptyset\}$. As a consequence, the complexities of the combinations are maximal for most cases. Then, each triplet combination of these bba are intended for the computation of Dubois & Prade rule. In particular, the entire triplet set are tested with different sizes, $K = N(N-1)(N-2)/6$:

N	20	40	60	80	100	120
K	1140	9880	34220	82160	161700	280840

and different computation approaches:

- Non recursive implementation with Spark,
- Recursive implementation with Spark,
- Non recursive implementation with Flink,
- Recursive implementation with Flink.

4.2   Results

Spark - Non recursive. The following table compiles the execution time for the *non recursive implementation with Spark*, for different numbers of threads and different triplet sets:

Spark - Non recursive

Threads	K	1140	9880	34220 (*)	82160 (*)	161700 (*)	280840 (*)
1	Time	6.4	56	235	740	2500	Failed
2		7.2	60	235	630	1500	3150
4		7.6	62	225	575	1350	2600
8		8.5	63	235	595	1260	2400
12		9.8	67	245	600	1270	2390
16		9.8	70	240	620	1300	2300

Postfix (*) indicates memory overflow and disk usage.

Spark - Recursive. The following table compiles the execution time for the *recursive implementation with Spark*, for different numbers of threads and different triplet sets:

Spark - Recursive

Threads	K	1140	9880	34220 (*)	82160 (*)	161700 (*)	280840 (*)
1	Time	5.1	42	161	550	1500	3400
2		5.4	44	155	430	1050	2280
4		5.8	45	160	385	940	1900
8		6.6	47	175	450	920	1800
12		7.2	51	180	480	980	1720
16		8	52	175	450	990	1800

Flink - Non recursive. The following table compiles the execution time for the *non recursive implementation with Flink*, for different numbers of threads and different triplet sets:

Flink - Non recursive

Threads	K	1140	9880	34220	82160 (*)	161700 (*)	280840 (*)
1	Time	17.5	128	415	980	1900	3490
2		12.8	83	240	630	1100	1970
4		10	49.7	145	440	775	1350
8		8.4	41.2	118	278	560	880
12		8.9	38	112	268	520	860
16		8.3	43	118	260	480	760

Flink - Recursive. The following table compiles the execution time for the *recursive implementation with Flink*, for different numbers of threads and different triplet sets:

Flink - Recursive

Threads	K	1140	9880	34220	82160 (∗)	161700 (∗)	280840 (∗)
1	Time	9.1	46	159	360	640	1220
2		7	31	93	249	430	750
4		6.1	19	55	173	310	500
8		6.2	16.8	45.8	115	230	400
12		6	16	44	133	270	380
16		5.8	18	41	120	240	340

4.3 Benefits and Limitations

The results confirm the efficiency of the recursive approach for *large* concomitant combination sequences. The recursive approach is however a burden for small sequences. Whatever, Flink outperforms Spark on sufficiently large cases, and shows especially good performances on recursive approach.

On this preliminary work, the code has not been optimized. For this reason, the table is not significant at this time in comparison with other existing optimized libraries. However, we compared the multithread implementations with the monothread implementations. The monothread recursive implementation is particularly efficient on sufficiently small processings. Whatever, it seems that for such small combination, a monothread implementation of the individual combination is better. This aspect is particularly visible with Spark implementation. In our example, Flink seems to behave better in the usage of the parallelism.

Moreover, these tests only considered the performance of simultaneous computation of large set of combinations, and especially, a full set of triple combinations. This implies important intermediate results caching. This case of use is then favorable to our recursive algorithm, since this approach reduces the caching by definition. But many other aspects of this implementation have to be investigated, in term of performance. The structure of the set of combinations may be used for optimizing the strategies of the computation flow. In that perspective, the capacity of Flink to implement computation streams and iterated processes will be particularly useful. The reactivity of this parallel computation on possibly complex single combinations is also a piece of performance to be evaluated precisely or optimized in the future, in regards to non-parallel approaches.

5 Conclusions

We proposed a generic distributed processing approach for computing belief combinations. The approach is based on a Map-Reduce paradigm, and has been implemented in Apache Spark and in Apache Flink. It is derived from the concept of referee function, introduced in a previous work with the aim of separating the definition of the combination rule from its actual implementation. This work has been completed by the proposal of a new recursive formalism for the definition of the rules, and of an improved Map-Reduce generic implementation. Some tests have been made for the rule of Dubois & Prade, which illustrated this computation improvement. This promising work will be extended in order to better take into account the computation flow structure. The reactivity of the computation and a better tuning of the parallelism level will be enhanced in the future for the purpose of an optimized library.

A Rules Definitions

A.1 Dubois & Prade Rule

The rule proposed by Dubois and Prade extends the conjunctive rule by redistributing disjunctively the conflict:

$$m_1 \oplus_{DP} m_2(X) = \sum_{Y_1, Y_2: \left\{ \begin{smallmatrix} Y_1 \cap Y_2 \neq \emptyset \\ Y_1 \cap Y_2 = X \end{smallmatrix} \right.} m_1(Y_1) m_2(Y_2) \tag{20}$$

$$+ \sum_{Y_1, Y_2: \left\{ \begin{smallmatrix} Y_1 \cap Y_2 = \emptyset \\ Y_1 \cup Y_2 = X \end{smallmatrix} \right.} m_1(Y_1) m_2(Y_2).$$

A.2 PCR6 Rule

The rule proposed by Martin and Osswald extends the conjunctive rule by a local proportional redistribution of the conflict:

$$m_1 \oplus_{PCR6} m_2(X) = \sum_{Y_1, Y_2: \left\{ \begin{smallmatrix} Y_1 \cap Y_2 \neq \emptyset \\ Y_1 \cap Y_2 = X \end{smallmatrix} \right.} m_1(Y_1) m_2(Y_2) \tag{21}$$

$$+ \sum_{Y: X \cap Y = \emptyset} \left[\frac{m_1(X)^2 m_2(Y)}{m_1(X) + m_2(Y)} + \frac{m_2(X)^2 m_1(Y)}{m_2(X) + m_1(Y)} \right].$$

References

1. Zhou, K., Martin, A., Pan, Q.: A similarity-based community detection method with multiple prototype representation. Physica A Stat. Mech. Appl. **438**, 519–531 (2015)
2. Zhou, K., Martin, A., Pan, Q., Liu, Z.: Median evidential c-means algorithm and its application to community detection. Knowl. Based Syst. **74**, 69–88 (2015)
3. Liu, W., Miller, P., Ma, J., Yan, W.: Challenges of distributed intelligent surveillance system with heterogenous information. In: Workshop on Quantitative Risk Analysis for Security Applications, Pasadena, California (2009)
4. Hong, X., Ma, W., Huang, Y., Miller, P., Liu, W., Zhou, H.: Evidence reasoning for event inference in smart transport video surveillance for video surveillance. In: 8th ACM/IEEE International Conference on Distributed Smart Cameras, Prague, Czech Republic (2014)
5. Dubois, D., Prade, H.: On the unicity of dempster rule of combination. Int. J. Intell. Syst. **1**, 133–142 (1986)

6. Lefevre, E., Colot, O., Vannoorenberghe, P.: Belief functions combination and conflict management. Inf. Fusion J. **3**, 149–162 (2002)
7. Smarandache, F., Dezert, J.: Information fusion based on new proportional conflict redistribution rules. In: International Conference on Information Fusion, Philadelphia, USA (2005)
8. Florea, M., Dezert, J., Valin, P., Smarandache, F., Jousselme, A.: Adaptative combination rule and proportional conflict redistribution rule for information fusion. In: COGnitive Systems with Interactive Sensors, Paris, France (2006)
9. Martin, A., Osswald, C.: Toward a combination rule to deal with partial conflict and specificity in belief functions theory. In: International Conference on Information Fusion, Qébec, Canada (2007)
10. Dambreville, F.: Definition of Evidence Fusion Rules Based on Referee Functions, vol. 3. American Research Press (2009)
11. Dambreville, F.: Map-reduce implementation of belief function rules. In: Proceedings of 5th International Conference on Data Management Technologies and Applications, Lisbon, Portugal (2016)
12. Dean, J., Ghemawat, S.: Mapreduce: simplified data processing on large clusters. Commun. ACM **51**, 107–113 (2008)
13. Zaharia, M., Chowdhury, M., Franklin, M., Shenker, S., Stoica, I.: Spark: cluster computing with working sets. In: Proceedings of 2nd USENIX Conference on Hot Topics in Cloud Computing, Berkeley, CA USA (2010)
14. Apache: Apache flink: scalable batch and stream data processing. https://flink.apache.org/
15. Dempster, A.P.: A generalization of bayesian inference. J. Roy. Stat. Soc. **B30**, 205–247 (1968)
16. Shafer, G.: A Mathematical Theory of Evidence. Princeton University Press, Princeton (1976)
17. Smets, P.: The combination of evidences in the transferable belief model. IEEE Trans. Pattern Anal. Mach. Intell. **12**, 447–458 (1990)

Log-Based Model to Enforce Data Consistency on Agnostic Fault-Tolerant Systems

Óscar Mortágua Pereira[⊠], David Apolinário Simões,
and Rui L. Aguiar

DETI, Instituto de Telecomunicações, University of Aveiro,
3810-193 Aveiro, Portugal
{omp, david. simoes, ruilaa}@ua. pt

Abstract. Agnostic fault-tolerant systems cannot recover to a consistent state if a failure/crash occurs during a transaction. By their nature, inconsistent states are very difficult to be treated and recovered into the previous consistent state. One of the most common fault tolerance mechanisms consists in logging the system state whenever a modification takes place, and recovering the system to the system previous consistent state in the event of a failure. This principle was used to design a general recovering log-based model capable of providing data consistency on agnostic fault-tolerant systems. Our proposal describes how a logging mechanism can recover a system to a consistent state, even if a set of actions of a transaction were interrupted mid-way, due to a server crash. Two approaches of implementing the logging system are presented: on local files and on memory in a remote fault-tolerant cluster. The implementation of a proof of concept resorted to a previous proposed framework, which provides common relational features to NoSQL database management systems. Among the missing features, the previous proposed framework used in the proof of concept, was not fault-tolerant.

Keywords: Fault tolerance · Logging mechanism · Software architecture · Transactional system

1 Introduction

Fault tolerance enables a system to keep its data consistent in the event of failure of some of its components [1]. A fault tolerant system either maintains its operating quality in case of failure or decreases it proportionally to the severity of the failure. On the other hand, a fault intolerant system completely breaks down with a small failure. Fault tolerance is particularly sought after in high-availability or life-critical systems.

Relational Database Management Systems (DBMS) are systems that usually enforce information consistency and provide atomic, consistent, isolated and durable (ACID) properties in transactions [2]. However, without any sort of fault-tolerance mechanism, both atomicity and consistency are not guaranteed in case of failure [3]. These ACID properties can be assured by a recovery system using a log-based model. Nevertheless, some critical aspects need to be addressed, such as: (1) logging processes are not simultaneous with the actions being executed, this way requiring a controlled

© Springer International Publishing AG 2017
C. Francalanci and M. Helfert (Eds.): DATA 2016, CCIS 737, pp. 141–159, 2017.
DOI: 10.1007/978-3-319-62911-7_8

supervision process on both the actions and the logging process; (2) the logging process must be fault-tolerant in order to be able to manage crashes during the logging process; (3) the recovery process must also be fault-tolerant in order to be able to manage crashes during the recovery process; (4) when cascading actions are implemented in the system, similar to the ones implemented by relational DBMS, those actions need also to be supervised.

We have proposed in a previous article a framework, hereafter referred to as Database Abstraction Framework (DFAF) [4], based in Call Level Interfaces (CLI) [5], that acts as an external layer and provides common relational features to NoSQL DBMS [5]. These features were ACID transactions, the execution of database functions (like stored procedures) and interactions with local memory structures, such as a ResultSet in JDBC [6] and a RecordSet in ODBC [7]. However, our framework lacked fault-tolerance mechanisms and, in case of failure, did not guarantee atomicity or consistency of information.

This paper presents a model that can be used to provide fault-tolerance to agnostic fault-tolerant systems through external layers. We describe how to log the system state, so that it is possible to recover and restore it when the system crashes; possible ways to store the state, either remotely or in the file system; and how to revert the state after a crash. As a proof of concept, the model was applied to our previous researches [8–12] and providing fault-tolerance to DFAF, which in turn provides ACID transactions to NoSQL DBMS. All the four aforementioned critical aspects are addressed by our model.

In other words, DFAF acts as an external layer over a DBMS, a deterministic system. A deterministic system is any process whose operations behave deterministically, and provide the same output with the same input. Non-deterministic events can happen in these systems and are either expected (e.g.: receiving a message), triggering deterministic behavior, or unexpected (e.g.: crashing), leading to undefined behavior.

The remainder of this paper is organized as follows. Section 2 describes common fault tolerance techniques and Sect. 3 presents the state of the art. Section 4 provides some context about the DFAF and Sect. 5 formalizes the proposed fault tolerance model, describing what information is stored and how to store it. Section 6 describes a fault-tolerant implementation. Section 7 presents the proof of concept. and, Sect. 8 evaluates our results and, finally, Sect. 9 presents our conclusions.

This paper is an extended version of a previous published paper in a conference proceedings [13].

2   Background

As previously stated, fault tolerance is a property of systems that do not stop as a whole due to hardware or software problems. A fault tolerant system remains operational, with an increased response time or a reduced throughput for example, in the event of a failure.

Fault tolerance is usually achieved by anticipating exceptional conditions and designing the system to cope with them. In [1], Randell et al. define an erroneous state as a state in which further processing, by the normal algorithms of the system, will lead to a failure [1]. When failures leave the system in an erroneous state, a roll-back mechanism must be used to set the system back in a safe state.

Techniques for handling failures can be classified into three categories [14]:

- Hardware Resilience: This category includes techniques implemented in processor, memory, storage and network hardware;
- Resilient Systems Software: This category includes software-based resilience techniques that are handled within systems software and programming infrastructure;
- Application-Based Resilience: The third category involves domain-specific models for fault tolerance that rely on information about the characteristics of the application (or the domain) to design specific algorithms that minimize the performance loss or system cost.

Hardware resilience is fully transparent to the user, but also beyond the scope of this paper. Application-based resilience, on the other hand, is a highly specialized category, and also falls beyond our scope. We will, as such, focus on Resilient Systems. In resilient systems, failures can be handled, depending on their severity, in three different levels [15]:

- Failure Masking: some failures can be hidden or their effect can be lessened (e.g.: lost messages can be retransmitted, resources can be replicated);
- Failure Recovery: software is designed in a way that the process state is periodically logged. When a failure occurs, processing is resumed from the last logged stated;
- Failure Tolerance: failures simply cannot be handled in an efficient manner, and the best choice is to inform a user or to abort the task (e.g.: informing a user that a server cannot be reached).

Regarding failure recovery, resilient systems rely on techniques like check-pointing, which is related to failure recovery, and process-, data- and task-driven techniques, which are specific to programming models and cannot be generalized. Check-pointing, however, is a popular and general technique that records the state of the system and, in the event of a fault, the system can be rolled-back and proceed from that point, instead of restarting completely.

Check-pointing in distributed systems encompasses uncoordinated, coordinated, communication-induced and log-based protocols [16]. In uncoordinated check-pointing [17], processes within a system take checkpoints independently, and in the event of failure, a consistent global state is found using dependency information from each checkpoint. In coordinated check-pointing [18], on the other hand, all individual checkpoints are part of a global consistent checkpoint. Communication-induced protocols force processes to take checkpoints based on messages from other processes. Log-based protocols require deterministic systems [19]. Non-deterministic events, such as the contents and order of incoming messages, are recorded and used to replay events that occurred since the previous checkpoint. Other non-deterministic events, such as hardware failures, are meant to be recovered from. Indirectly, they are recorded as *lack of information*.

Check-pointing can also be done with complete or incremental checkpoints. A complete system-level checkpoint saves the entire address space of a process. On the other hand, incremental checkpoints only save modified parts of the address space, in order to reduce the size of a checkpoint.

3 Related Work

There are many logging system proposals from the scientific community [20–23] whose purpose is not to provide fault tolerance, but simply to store information which is later used in data analytics or statistical analysis. In the fault tolerance context, there are proposals like Ralph et al.'s system [24], which scans a large number of variables and arranges the data to be reviewed by an operator, so that abnormal process is detected; or popular write-ahead logging approaches [25], commonly used by DBMS to guarantee both atomicity and durability in ACID transactions.

There are also other approaches which do not rely on logging systems to provide fault tolerance, like Huang et al.'s method and schemes for error detection and correction in matrix operations [26]; Rabin et al.'s algorithm to efficiently and reliable transmit information in a network [27]; or Hadoop's data replication approach for reliability in highly distributed file systems [28]. Some relational DBMS use shadow paging techniques [29] to provide the ACID properties. However, all of the above described fault tolerance mechanisms are specific to a given system. While they follow a broader model (for example, Hadoop's approach is based in data replication), most proposals are adapted to a particular context and integrated with existing solutions, and therefore not suitable to be used in an external fault-tolerance layer.

In the category of data replication, the scientific community has also proposed several algorithms and mechanisms, such as [30], which is based on a primary copy technique; [31], an LDAP-based replication mechanism; or [32], which provides an adaptive algorithm that replicates information based on its access pattern. Recently, proposals have also taken into account byzantine failure tolerance [33–38]. Byzantine fault-tolerant algorithms have been considered increasingly important because malicious attacks and software errors can cause faulty nodes to exhibit arbitrary behavior. However, the byzantine assumption requires a much more complex protocol with cryptographic authentication, an extra pre-prepare phase, and a different set of techniques to reach consensus.

To the best of our knowledge, there has not been work done with the goal of defining a general logging model that provides fault tolerance as an external layer to an underlying deterministic system. Some solutions provide fault tolerance, but are adapted to a specific context. Others are overly-abstract general models, like data replication, and do not cover how to generate the necessary said data from an external layer to provide fault-tolerance to the underlying system. Not only that, but many data replication systems also assume conditions we do not, such as the possibility of byzantine failures, or overly complex data access patterns. While byzantine failures are of enormous importance in distributed unsafe systems, such as in the BitCoin [39] environment, we consider their countermeasures to be complex and performance-hindering in the scope of our research. Not only that, but byzantine assumptions have been proven to allow only up to 1/3 of the nodes to be faulty.

We intend to focus on fault-tolerance for underlying deterministic systems through a logging system, and while distributed data replication is used for reliability, expected DFAF use cases do not assume malicious attacks to tamper with the network.

However, our model is general enough that it supports the use of any data replication techniques to replicate logging information across several machines.

4 Context

We have previously mentioned the DFAF, which allows a system architect to simulate non-existent features on the underlying DBMS for client applications to use, transparently to them. Our framework acts as a layer that interacts with the underlying DBMS and with clients, which do not access the DBMS directly. It allowed ACID transactions, among other features, on NoSQL DBMS, but was not fault tolerant.

Typically, NoSQL DBMS provide no support to ACID transactions. NoSQL data stores are sometimes referred as Basically Available, Soft state, and Eventually consistent (BASE) systems [40]. In this acronym, Basically Available means that the data store is available all the time whenever it is accessed, even if parts of it are unavailable; Soft-state highlights that it does not need to be consistent always and can tolerate inconsistency for a certain time period; and Eventually consistent emphasizes that after a certain time period, the data store comes to a consistent state.

An ACID transaction allows a database system user to arrange a sequence of interactions with the database which will be treated as atomic, in order to maintain the desired consistency constraints. For reasons of performance, transactions are usually executed concurrently, so *atomicity*, *consistency* and *isolation* can be provided by file- or record-locking strategies.

Transactions are also a way to prevent hardware failures from putting a database in an inconsistent state. DBMS usually have mechanisms for single-statement transactions, but our framework must be adjusted to take hardware failures into account with multi-statement transactions. In a failure free execution, our framework registers which actions are being executed in the DBMS and how to reverse them. Actions are executed in the DBMS immediately and are undone if the transaction is rolled-back.

However, during a DFAF server crash, the ACID properties are not enforced. As an example, consider a transaction with two insert statements. If the DFAF server crashed after the first insert, even though the client had not committed the transaction, that value would remain in the database, which would mean the *atomic* aspect of the transaction was not being enforced.

To enforce it, we designed a logging mechanism, whose records are stored somewhere deemed safe from hardware crashes. That logging system will keep track of the transactions occurring at all times and what actions have been performed so far. When a hardware crash occurs, the logging system is verified and interrupted transactions are rolled-back before the system comes back on-line and new transactions are executed.

Our logging system is a log-based protocol where the underlying DBMS acts as the deterministic system mentioned previously. Each action in a transaction represents a non-deterministic event and is, as such, recorded, so that the chain of events can be recreated and undone when the system is recovering from failure, as if the transaction had simply been rolled-back and had never crashed.

5   Local Architectural Model

In this chapter, the local conceptual model is presented. It is based on a mechanism responsible for ensuring that the last consistent system state is reached if a crash occurs during a transaction. Two implementations will be provided of the conceptual model: one local and other remote

Logging systems for fault-tolerance mechanisms have several different aspects that need to be defined. The main key issue, is that the logging process must itself be fault-tolerant. The situations to be addressed are:

First of all: the logging system must be designed in a way that the logging is not affected by hardware or software failures, particularly during the logging process. In other words, if the server crashes while a database state is being logged, the system must be able to handle an incomplete log and must be able to recover its previous state.

Second of all: logging actions is not done at the same time as the actions are executed. Taking an insertion in a database as an example, the system logs that a value is going to be inserted, the value is inserted and the system logs that the insertion is over. However, if the system crashes between both log commands, there is no record of whether the insert took place or not. To solve this, the underlying system must be analyzed to check if it matches the state prior to the insertion or not.

Thirdly: while recovering from a failure, the server can crash again, which means the recovery system must also be fault tolerant.

Finally: cascading actions imply multiple states of the underlying system, all of which must be logged so that they can all be rolled-back. In other words, if an insert in a database triggers automatically an update, then the database has three states to be logged: the initial state, the state with the insertion and the state with the insertion and the update. Because the server can crash at any of these states, they all need to be logged so that the recovery process rolls-back all the states and nothing more than those states.

If all of these aspects are taken into account, our logging mechanism provides fault tolerance to underlying deterministic systems.

5.1   Key Concepts

The logging process is aimed at dealing with the transactional concept. Basically, there is a trigger responsible for starting the execution of a set of operations and also a trigger to end the process. If the system crashes between both, there is the need to roll-back the system's state to the previous consistent state. The logging process is based on logging information whenever the state of the deterministic system is changed.

The logging information stored will inherently have the actions being performed in the underlying system. In order to provide fault tolerance, some actions will undoubtedly need to be reversed during a recovery process, to avoid leaving the system in an inconsistent state. As such, along with the actions performed, the system must also log how to undo them. In other words, when a client issues an action, the action to revert it, hereafter referred to as the *reverser*, is automatically calculated. For example, in a database, an insert statement's reverser is a delete statement. Reversers are executed backwards in a recovery process to keep the underlying system in a consistent state.

However, logging actions and performing them cannot be done at the same time. It is also not adequate to log an action after it has already been performed, since the server could crash between both stages (executing and logging the action), and there would be no record that the system state had changed. Therefore, actions (and their reversers) must be logged before they are executed on the underlying system. This, however, can still lead to problems, if the server crashes between the log and the execution, since the recovery process would try to reverse an action that had not been executed.

To solve this problem, the logging system can also record that an action has been completed, but the problem remains if the server crashes between the execution and the logging. Because we have no assumptions regarding when the system can crash, the only way to solve this problem is to directly assess the underlying system's state to figure out whether the action has been performed or not. Since we have access to the underlying system's state prior to the action being executed, we can find a condition that describes whether the action has been executed or not. This condition is hereafter referred to as the *verifier*. It must be here emphasized that the logging process is independent from the system being protected. Additionally, it is our intention to design a model that does not depend on any property of the system being protected. For example, the logging system could resort to SQL triggers in case they were supported. Unfortunately, these are only supported by a small cluster of databases based on the SQL standard. In spite the possibility of being used with a great success (triggers), a general model, as the one herein presented, cannot rely on such an approach.

Consider the insertion of a row with value 'A' in a DBMS. The insertion of this value can be verified by the amount of rows with value 'A' that existed prior to the insertion. If there were two 'A's and the transaction crashed during the insertion of a third, by counting how many exist in the database, the necessity of reverting this action can be inferred. It would be necessary if there are now three 'A's and unnecessary if there are still two 'A's.

The concept is extended to cascading actions. A reverser is computed for each cascading action, as well as a verifier to determine whether this effect happened and needs to be rolled-back or not. Consider that, in the previous example, each insertion triggered an update on another table that counted how many 'A's existed in the table. The logging information will contain the action desired by the user (insertion of 'A'), the reversers (deletion of 'A' and update of the count number) and the verifiers (there were two 'A's in the database and the count value showed the number 2). If the server crashes during these triggered actions or during a rollback, each condition must be checked before applying each reverser, to make sure the same action is not reverted twice or that an action that was not executed is not reverted.

During the recovery process, all reversers are executed backwards, but only if the reverser's corresponding verifier shows the need to execute it. If it does not, it is simply removed from the log file. After an action has been reversed, its record (along with the reverser and verifier) is removed from the log. If the server crashes during a recovery, due the verifier system, there is no risk of reverting actions that need not be reverted or that have not yet been executed.

5.2   Logging Model

The logging model can be based on files where the relevant information is stored: performed database actions, underlying database system state and reversers. These files need to be judiciously observed, regarding its state and its contents. Three files are used to this end: Log file is used to keep the performed actions, the underlying system state and the reversers; the Copy file is used to keep the contents of the Log file while the Log file is being updated for the next action; the Temp file is used as a flag in order to know if the Copy file has been created successfully. A more detailed description is provided below, including a step by step description of the critical aspects to be addressed. Figure 1 presents the state diagram for the logging system model. Each state is identified by a code that comprises two letters shown between curved brackets.

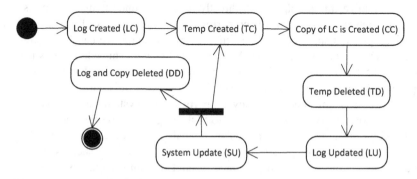

Fig. 1. State diagram for the logging system model.

A new empty Log file is created for each new transaction occurring in the system (LC). The file is created before the first action of every new transaction and it is deleted just after the transaction is completed (DD). If the system crashes when the transaction is starting and creating the Log file, the file can either exist and be empty or not exist. In case it exists, it contains no actions to be rolled-back, thus the Log file is deleted. If the server crashes when deleting the file and closing the transaction, the file can either exist with its contents still intact or not exist. If it does not exist, the transaction was already over. If it still exists, then it is possible to read it and rollback until the previous stable database state and, finally delete the Log file.

The logging process must be done in a way that the logging system's last state must be recoverable. As such, to prevent state corruption, we must always keep a copy of the old state (Log file) until the new one is completely defined. To achieve this result, as previously mentioned, two additional files are used: *Temp* and *Copy*. While the Log file is used to keep track of the history (actions, states and reversers) of all actions already performed and also the one under execution (if it is the case), the Temp file is used to indicate that a copy of the Log file is in process, which means that the contents of the Copy file cannot be trusted. Thus, for each new action, a new empty Temp is created (TC), and the Log file's content is copied to a new Copy file (CC). After copying the Log file, the Temp file will be deleted (TD). If the system crashes before deleting the Temp file, it means that the content of the Copy file cannot be trusted but the content of the Log file can be trusted and used to roll-back the transaction. As previously mentioned, if the content of Log file is successfully copied to the Copy file then the Temp file is deleted (TD), meaning that the content of Copy file can be trusted and the content of Log file cannot be trusted anymore. The content of Log file cannot be trusted because after deleting the Temp file, the updating process of the Log file (LU) will be in process. During this process, the system can crash leading to an uncertainty about the validity of the content of the Log file. Thus, in this situation, to recover the system's previous state, the content of the Copy file is used.

Table 1 shows the several phases described above, the phase of each of the files, and what file is chosen on each phase. The meaning of each symbol is: 'o' – file exists and its content is reliable; 'x' – file does not exist; '?' – the file exists but its content is not reliable. The table content is now described. On phase A, there is no action being performed. Two situations can trigger this: (1) a new transaction has started and the Log file is created as an empty file; (2) a transaction is taking place but no action is being processed. In the latter situation, an action has just been executed but the next one (if it exists) did not start yet. Phase B is triggered when a new action is required to be executed: Temp file is created. During phase C, the Copy file is created and the content of the Log file is copied. If the server crashes on phases A, B or C, the chosen recovery file is the original one – the Log file. Despite a new action has been triggered, the content of the Log file is reliable because it has not been updated during these phases. On phase C, Temp file is deleted, meaning that the content of the Log file has been successfully copied to the Copy file and, consequently, the content of the Copy file is now reliable. Phase D, derives immediately from phase C. After having deleted the Temp file, it is not possible to know if the content of the Log file is reliable, since the system will start the update process of the Log file, so that it includes the required information to roll back the action being processed. If the server crashes on phases C or D, the chosen file is the Copy file. Finally, on phase E, the action being processed has been successfully executed, the Log file successfully updated and, therefore, the Copy file can be deleted. In case the system crashes, the chosen recovery file is the Log file.

Table 1. A log-update cycle.

	Phases				
	A	B	C	D	E
Log	o	o	o	?	o
Temp	x	o	o	x	x
Copy	x	x	?	o	x
Valid file	Log	Log	Copy	Copy	Log

The model just presented is used in the local storage mechanism. All actions in the transaction trigger this update, and all reverts during the recovery process do so as well. Even if the server crashed during multiple concurrent transactions or during a recovery process, all uncommitted transactions will be logged and this allows the system to roll them back and to leave the database on a consistent state.

As shown in Sect. 8, the local architectural model relies on writing the logging information to disk: fault tolerance is supported even in a complete system crash, but with heavy performance costs when transactions have many operations. The performance costs are basically justified by I/O disk operations. The Log, Temp and Copy files need to be persistent in order to ensure that, if the system crashes, the log information is not lost.

We assume that the hardware crashes will not be so severe that they render the hard drive contents unrecoverable or that a back-up system is deployed to allow the recovery of a defective file system. Most file systems do not provide fault-tolerant atomic file creation, removal, copy, movement, appending or writing operations, which is the reason behind the protocol shown in Fig. 1.

6   Remote Architectural Model

The remote architectural model tries to leverage both performance and fault tolerance and relies on a remote machine to keep the logging information in memory. I/O operations are not as heavy on performance as writing to disk, but fault tolerance is only guaranteed if the logging server does not crash. In order to overcome this fragility, a fault-tolerant master-slave architecture was designed, hereafter referred to as a Cluster Network (CN). CN allows several machines to coordinate and replicate information among them. This system can be used to store the logs from the remote mechanism, which allows some machines to crash without loss of information. In the designed algorithm, the only case where the logs would be lost would be a scenario where all machines crashed, which is unlikely if the machines are geographically spread. We expect the performance of this mechanism to be superior against the local architectural model.

The remote architectural model uses TCP sockets to exchange information between the servers. Because TCP provides reliability and error control, both machines know when a message has been properly delivered and the system server can perform the requested actions while the logging server keeps the information in memory. Both servers can detect if the network failed or the remaining server has crashed. In these cases, the recovery process can be initiated until connectivity is re-established. When using a CN to store the needed information, the remote mechanism allows for fast interactions (faster than, for example, writing information to the file system), reliability (information is replicated through the nodes in the cluster to avoid losses if some of the machines crash) and consistency (the nodes all have the same information). Data replication techniques such as the byzantine tolerant approaches mentioned previously are a valid option, but have an associated performance decay due to the byzantine assumption and a low threshold for the amount of faulty machines. As such, the CN was designed as a fault-tolerance master-slave network that replicates information across all the slaves and better fits our requirements.

We require our CN to be able to scale as needed, without having to interrupt service or without having maintenance downtime. We considered that nodes should be symmetrical to avoid the human error factor present in id-based systems. We also want a stable algorithm (a master node remains a master node until it crashes) to avoid unnecessary operations when a former master node is turned back on. Finally, we consider that an IP network is not perfect and that network elements (switches, routers) and well as network links can crash at any time.

We therefore define our CN as a set of at least one node that communicates through IP, where any of the nodes can crash and be restarted at any given time. The master node is contacted by clients and it forwards the information to the slave nodes. Clients can find the master node through any number of methods, like DNS requests, manual configuration, broadcast inquiries, etc. If the master crashes, one of the slaves is nominated to be master and, because all the information was replicated among the slaves, it can resume the master's process.

Our leader-election algorithm is inspired by Gusella et al.'s election algorithm [41]. While many other leader election algorithms would be supported, this one suits the DFAF requirements the best. The authors have developed a Leader Election algorithm that is dynamic (nodes can crash and restart at any time), symmetric (randomization is used to differ between nodes), stable (no leader is elected unless there is no leader in the cluster) and that uses User Datagram Protocol (UDP) communication (non-reliable, non-ordered). It supports dynamic topology changes to some degree, but it is not self-stabilizing (nodes start in a defined state, not in an arbitrary one).

When a master is defined, the master is the one receiving requests from clients. In order to guarantee consistency among all the nodes, the master forwards any incoming requests to the slaves before answering the client with the corresponding response. This guarantees that all the slaves will have the same information as the master. If the master crashes during this process, because the client still has not been answered, he will retry the request to the new master, which will store it (while avoiding request duplication) and forward it to the slaves.

When a slave joins the network, he contacts the master and requests the current system information (in this case, the current log). A mutual exclusion mechanism is necessary to avoid information inconsistency when information is being relayed to a new slave (if new information reaches the master while a new slave is not fully updated). To avoid request duplication from clients when the master node crashes, a request identification number is used.

Using this approach means that up to $N - 1$ nodes in a CN with N nodes can crash without information being lost or corrupted. Using other approaches for data replication, such as [33], only allows up to $N/3$ nodes to be faulty and is expected to have worse performance. However, byzantine-tolerant approaches are more robust and our logging model is general enough that any data replication mechanism can be used to safe-keep the logging information.

7 Proof of Concept

In this chapter, two proofs of concept are presented, each one for one of the two architectural models. The logging mechanism was adapted to the previously mentioned DFAF, in order to guarantee the *atomic* and *consistent* properties of transactions. This way, even if the DFAF server crashed during multiple concurrent transactions, those transactions will all be rolled-back and the underlying database will be on a consistent state when the recovery process has finished.

That recovery process may activate manually or automatically, either after the crash or after a server restart. If the recovering process is based on the local architectural model (information stored on the local hard drive), it can only start after the server has been restarted. If the recovering process is based on the remote architectural model (information is stored remotely), the recovery process can start when a system administrator notices that the server has crashed. If a remote server is being used to store the information, it can detect automatically that the server has crashed (through a time-out system, for example) and it can start the process automatically.

The reverser and verifier system in DFAF depends on the underlying DBMS schema and query language. Different schemas can imply different cascading actions, if, for example, different triggers are defined in each schema. Different query languages also imply different reversers and verifiers, since an insert in SQL has a very different syntax from a NoSQL DBMS with a custom query language.

Having multiple transactions occurring at the same time implies having either multiple log files or a single log file with information from all transactions. This could lead to problems during the recovery process, if the order of actions in separate transaction was not being logged. However, the fact that transactions guarantee the isolation property means that each of their actions will not affect other transactions. Therefore, the order in which each transaction is rolled-back is irrelevant, as long as the statements in each transaction are executed backwards.

To prove our concept, the local logging mechanism using DFAF was tested with a single client connecting to the database. The client starts a transaction, inserts a value and updates that value, finishing the transaction. During this process, the logging information is stored in a local file, as can be seen in Listing 1.

```
Start DFAF transaction
    Connect to database
    Create empty 'log' file
Insert value "A"
    Create 'temp' file
    Back-up 'log' file as 'copy' file
    Delete 'temp' file
    Update 'log' with action, reverser and verifier
    for 'insert A'
    Delete 'copy' file
    Execute action in database
Update value "A" for "B"
    Create 'temp' file
    Back-up 'log' file as 'copy' file
    Delete 'temp' file
    Update 'log' with action, reverser and verifier
    for 'update A to B'
    Delete 'copy' file
    Execute action in database
Close DFAF transaction
    Delete 'log' file
    Close connection
```

List. 1. Pseudo-code example of a transaction.

We crashed the transaction on several stages of each action's execution and verified that the recovery process could correctly interpret the correct log file and set the database in a correct state, the one previous to the transaction. In order to interrupt the process on particular stages, exceptions were purposely induced in the code, which were thrown at the appropriate moments. The recovery process was then started and tested as to whether it could successfully recover and interpret logged information and, if needed, rollback the database to a previous state. An example of a recovery process is shown in Listing 2.

```
If 'log' files exist
    For each 'log' file
        Pick appropriate file to read according to
            Table 1
        Remove additional files
        Set valid file as 'log' file
        For each action in the log (backwards)
            If verifier shows action was performed
                Execute reverser
            Create 'temp' file
            Back-up 'log' file as 'copy' file
            Delete 'temp' file
            Remove action, reverser and verifier from
            'log' file
            Delete 'copy' file
        Delete 'log' file
Accept incoming clients/transactions
```

List. 2. Pseudo-code example of a recovery process.

Results showed that the system was able to recover from a failed transaction and returned the database to a safe state in all cases.

To prove our concept with a CN, or in other words, with the remote mechanism, we deployed a network with a client connected to a DBMS and to a CN, as shown in Fig. 2.

We used the same transaction used to test the local mechanism. In our first test, we checked whether the CN could detect and roll-back failed transactions. We crashed the client after the first insertion and the CN detected the crash and rolled-back the transaction. An example of this process, from the master-node's perspective, is shown in Listing 3.

```
If client timed-out
    For each logged action (backwards)
        Inform slaves that action is about to be reverted
            If verifier shows action was performed
                Execute reverser
            Remove action, reverser and verifier from log
            Confirm to slaves that action was reverted
    Clear client information
```

List. 3. Pseudo-code example of a master node in a recovery process.

Fig. 2. The deployed network for test with the remote mechanism and single client, from [13].

In our second test, we checked if a correct rollback was ensued with crashes on different stages of the transaction (before logging the action, after logging but before performing the action, after performing but before logging that it has been performed and after logging that the action had been done), and monitored the roll-back procedure to guarantee the database was in the correct state after the recovery process had finished.

Finally, we checked whether several concurrent transactions occurring in a DFAF server could all be rolled-back without concurrency issues. We used a DFAF server to handle several clients while connected to a CN, as can be seen in Fig. 3, and crashed the server during the client's transactions. The CN detected the crash and rolled-back all transactions, leaving the database once more in a consistent state.

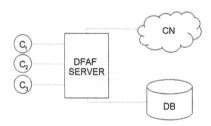

Fig. 3. The deployed network for tests with the remote mechanism and multiple clients, from [13].

8   Evaluation

To demonstrate the soundness of the presented approach in a practical environment, the performance of our logging mechanism's implementation was examined. The test-bed used a 64-bit Linux Mint 17.1 with an Intel i5-4210U @ 1.70 GHz, 8 GB of RAM and a Solid State Drive. For tests involving a CN, a second machine was used, running 64-bit Windows 7 with an Intel i7 Q720 @ 1.60 GHz, 8 GB of RAM and a Hard Disk Drive. A 100 Mbit direct and dedicated cable network was used as an underlying communication system between both nodes.

Figure 4 shows how the local (green) and remote (red) logging mechanisms compare with each other, using as a basis for comparison a transaction with up to one to one thousand (1,000) statements on a SQLite table. This number of statements was based on previous DFAF evaluations. Tests were repeated at least 20 times to get an average of the values. The 95% confidence interval was calculated, and the base time for operations was removed to allow for a more intuitive graph analysis. The CN used

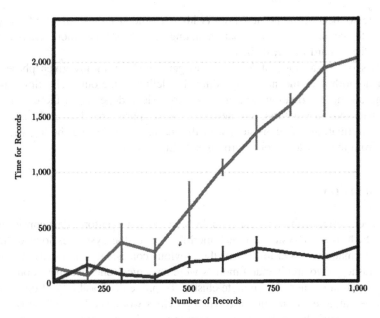

Fig. 4. Performance (in milliseconds) of the different logging mechanisms, from [13] (Color figure online).

for the remote mechanism was a local single-node, which removed most of the network interference with the tests.

In a first view, the graphic shows that the behavior of the local and the remote architectures diverge as the number of actions increases in a transaction. Although, it is also noticeable that until 250 actions both architectures present similar response times. If a closer look is taken, the local architecture presents a better response time for small numbers of actions (around 100). The reason for this result, is that the overhead induced by the network clearly overrides the I/O operations for the Log, Temp and Copy files. As expected, for transactions with more actions, the most performant mechanism is the remote mechanism, where a sub-second performance decay is noticed (around 321 ± 209 ms for 1000 operations). The baseline time for 1,000 operations was $10,295 \pm 1,142$ ms, which means remote mechanism has a performance decay of approximately 3.1%. The local mechanism is the least performant, due to the high amount of disk operations, with around $2,047 \pm 237$ ms for 1,000 operations, a 19.8% performance decay. The performance difference of an order of magnitude between both mechanisms is due to the fact that, as the logging file gets bigger, it takes longer to read, copy and write it. This means that, with a transaction of 1.000 insertions, for example, the last insertion will take a lot longer than the first insertion, while the remote mechanism takes a constant amount of time for any insertion.

We tested Cluster Networks to find how long it takes to find a master and make the information consistent among them. These values have a direct correlation to the defined time-outs on each state of the network, as defined by Gusella et al.'s algorithm. We created two-node networks (1 master, 1 slave) and measured the times taken for each node to become a master/slave (with a confidence interval of 95%) and to guarantee the consistency of information among them. Tests with more nodes were not feasible, due to hardware restraints.

Tests show an average of 5 ± 1 ms to get a node from any given phase of the election algorithm to the next, excluding the defined time-outs. The time taken to exchange all the information from a master to a slave depends on the current information state, but in our tests, any new slave took approximately 8 ± 1 ms to check whether information was consistent with the master. Transferring the log with 1,000 records from the first test took approximately 20 ± 4 ms.

9 Conclusion

In this paper, a recovering log-base model was presented to enforce data consistency on agnostic fault-tolerant systems. Every modification to the system is logged and this logged information is used to recover the previous consistent state in case the system fails or crashes. Two architectural models were presented: one local based on files and another based on memory in a remote cluster. To prove the feasibility of our proposal, it was tested in an environment based on a previous work, the DFAF. The DFAF is a CLI-based framework that implements common relational features on any underlying DBMS that do not support them. These features can be very useful on situations where NoSQL are being used. Some of the features include ACID transactions, local memory structure operations and database-stored functions, like Stored Procedures. However, the proposal lacked a fault tolerance mechanism to ensure the atomic property of transactions in case of failure. We now propose a fault tolerance model, general enough to work with underlying deterministic systems, and, in this paper, adapted to DFAF. Our model is a logging mechanism which requires the performed action, its verifier (that checks whether it has been executed or not) and its reverser (to undo it, in case of failure). Two ways of storing the information were presented: either locally in the file system, or remotely in memory in a dedicated cluster. Because operating systems do not usually provide atomic operations, to prevent the logging information from becoming corrupted, a description was made of how to update the information in the logging model. In order to guarantee that the model is also fault tolerant and the information is not lost in case of failure, a description of a master-slave network was also presented that can be used to replicate the information. Clients contact the master, which replicates the information to slaves without consistency issues.

Our performance results show that the use of our logging mechanism can be suitable for a real-life scenario. There is an expected performance degradation, but a fault tolerant systems provide several advantages over a slightly more performant not fault-tolerant systems. The performance of both architectures clearly diverges depending on the context (number of actions). If the number of actions is under 100, the local architecture presents a better performance. But if the number of actions is higher than 100, then the remote architecture presents a better performance. While the performance of the local architecture decays are the number of actions increase, the performance of the remote architecture is mostly only very slightly depends on the number of actions However, if a closed look it taken to the collected results, the overhead on transactions with tenths of actions is almost negligible, in either implementations: local and remote depending on the number of actions. By empirical evidence, we emphasize that these are most common situations.

In the future, the local and remote mechanisms will be improved. Regarding the file system, a highly performant algorithm will be developed, that does not rely on copying the previous log on each update. Regarding the remote mechanism, the CN will be adapted for other requirements, in order to improve performance. This can be done by allowing priority nodes and removing the symmetry factor. This way, servers can preferentially become masters, if they have better hardware or conditions. The CN can also be improved by changing the underlying communication protocol, which at the moment is assumed to be unreliable. A master look-up mechanism will also be developed, like DNS registration. At the moment, there is no such mechanism, and clients resort to finding masters manually. Using DNS registration, clients can simply use the DNS look-up system to find masters when there is a crash.

Summarizing, a logging-based model with DFAF was implemented, this way guaranteeing the atomic property of transactions on an underlying DBMS, even in cases of failure. Two ways of storing the information were provided, to leverage performance and reliability. A master-slave fault tolerant network was also proposed, which can be used as a remote server to keep information replicated and consistent. Both the logging model and the CN can be used for other applications; we have for example adapted the CN to act as a concurrency handler in another module of DFAF. In real scenarios, there is the need to choose which of the architectures better fits the necessary requirements.

Acknowledgements. This work is funded by National Funds through FCT – Fundação para a Ciência e a Tecnologia under the project UID/EEA/50008/2013.

References

1. Randell, B., Lee, P., Treleaven, P.C.: Reliability issues in computing system design. ACM Comput. Surv. **10**, 123–165 (1978). doi:10.1145/356725.356729
2. Sumathi, S., Esakkirajan, S.: Fundamentals of Relational Database Management Systems. Springer, Heidelberg (2007). doi:10.1007/978-3-540-48399-1
3. Gray, J.: The transaction concept: virtues and limitations. In: Proceedings of 7th International Conference on Very Large Data Bases, pp. 144–154 (1981). doi:10.1.1.59.5051
4. Pereira, Ó.M., Simões, D., Aguiar, R.L.: Endowing NoSQL DBMS with SQL features through standard call level interfaces. In: SEKE 2015 - International Conference on Software Engineering Knowledge Engineering, pp. 201–207 (2015)
5. ISO: ISO/IEC 9075-3:2003 (2003). http://www.iso.org/iso/catalogue_detail.htm?csnumber=34134
6. Parsian, M.: JDBC Recipes: A Problem-Solution Approach. Apress, New York (2005)
7. Microsoft RecordSet (ODBC): Microsoft. http://msdn.microsoft.com/en-us/library/5sbfs6f1.aspx. Accessed 16 Nov 2016

8. Pereira, Ó.M., Aguiar, R.L., Santos, M.Y.: CRUD-DOM: a model for bridging the gap between the object-oriented and the relational paradigms. In: ICSEA 2010 - International Conference on Software Engineering Applications, Nice, France, pp. 114–122 (2010)

9. Pereira, Ó.M., Aguiar, R.L., Santos, M.Y.: An adaptable business component based on pre-defined business interfaces. In: 6th ENASE Evaluation of Novel Approaches to Software Engineering, Beijing, China, pp. 92–103 (2011)

10. Pereira, O.M., Aguiar, R.L., Santos, M.Y.: ABC architecture: a new approach to build reusable and adaptable business tier components based on static business interfaces. In: Maciaszek, L.A., Zhang, K. (eds.) ENASE 2011. CCIS, vol. 275, pp. 114–129. Springer, Heidelberg (2013). doi:10.1007/978-3-642-32341-6_8

11. Pereira, Ó.M., Regateiro, D.D., Aguiar, R.L.: Secure, dynamic and distributed access control stack for database applications. In: SEKE 2015 - International Conference Software Engineering Knowledge Engineering, pp. 365–369 (2015)

12. Pereira, Ó.M., Regateiro, D.D., Aguiar, R.L.: Secure, dynamic and distributed access control stack for database applications. Int. J. Softw. Eng. Knowl. Eng. 25, 1703–1708 (2015). doi:10.1142/S0218194015710035

13. Pereira, Ó.M., Simões, D.A., Aguiar, R.L.: Fault tolerance logging-based model for deterministic systems. In: DATA 2016 - 5th International Conference on Data Science Technology Application, Lisbon, Portugal, pp. 119–126. SCITEPRESS (2016)

14. Balaji, P., Buntinas, D., Kimpe, D.: Fault Tolerance Techniques for Scalable Computing. In: McsAnlGov, pp. 1–33 (2012)

15. Borges, A.R.: Introductory concepts. Lecture on Distributed Systems (2015)

16. Elnozahy, E.N.(M)., Alvisi, L., Wang, Y.-M., Johnson, D.B.: A survey of rollback-recovery protocols in message-passing systems. ACM Comput. Surv. 34, 375–408 (2002). doi:10.1145/568522.568525

17. Bhargava, B., Lian, S.-R.: Independent checkpointing and concurrent rollback for recovery in distributed systems - an optimistic approach. In: Proceedings of Seventh Symposium Reliable Distributed Systems, pp. 3–12 (1988). doi:10.1109/RELDIS.1988.25775

18. Chandy, K.M., Lamport, L.: Distributed snapshots: determining global states of distributed systems. ACM Trans. Comput. Syst. 3, 63–75 (1985). doi:10.1145/214451.214456

19. Johnson, D.B.: Distributed system fault tolerance using message logging and checkpointing. Ph.D. Dissertation. Rice University, Houston, TX, USA. AAI9110983 (1990)

20. Brown, M.: Event logging system and method for logging events in a network system (1999)

21. Heemels, J.P., Carlson, G.M., Spinelli, J.C.: Data logging system for implantable cardiac device (1997)

22. Fraker, W.F., Storm, J.M.: Position and time-at-position logging system (1999)

23. Salmassy, O.E., Sullivan, R.E.: Statistical and environmental data logging system for data processing storage subsystem (1972)

24. Anderson, R.A.: Automatic process logging system (1959)

25. Mohan, C., Haderle, D., Lindsay, B., et al.: ARIES: a transaction recovery method supporting fine-granularity locking and partial rollbacks using write-ahead logging. ACM Trans. Database Syst. 17, 94–162 (1992). doi:10.1145/128765.128770

26. Huang, K.H., Abraham, J.A.: Algorithm-based fault tolerance for matrix operations. IEEE Trans. Comput. 33, 518–528 (1984). doi:10.1109/TC.1984.1676475

27. Rabin, M.O.: Efficient dispersal of information for security, load balancing, and fault tolerance. J. ACM 36, 335–348 (1989). doi:10.1145/62044.62050

28. Shvachko, K., Kuang, H., Radia, S., Chansler, R.: The Hadoop distributed file system. In: 2010 IEEE 26th Symposium on Mass Storage Systems and Technologies MSST2010 (2010). doi:10.1109/MSST.2010.5496972

29. Ylönen, T.: Concurrent shadow paging: a new direction for database research (1992)

30. Oki, B.M., Liskov, B.H.: Viewstamped replication: a new primary copy method to support highly-available distributed systems. In: PODC 1988 Proceedings of Seventh Annual ACM Symposium on Principles of Distributed Computing, vol. 62, pp. 8–17 (1988). doi:10.1145/62546.62549

31. Shih, K.-Y., Srinivasan, U.: Method and system for data replication (2003)

32. Wolfson, O., Jajodia, S., Huang, Y.: An adaptive data replication algorithm. ACM Trans. Database Syst. **22**, 255 (1997)

33. Castro, M., Liskov, B.: Practical Byzantine fault tolerance. In: Proceedings of Symposium on Operating Systems Design and Implementation, pp. 1–14 (1999). doi:10.1145/571637.571640

34. Bershad, B., ACM Digital Library, ACM Special Interest Group in Operating Systems, et al.: HQ replication: a hybrid quorum protocol for Byzantine fault tolerance. In: Proceedings of 7th Symposium on Operating System Design and Implementation, p. 407 (2006)

35. Castro, M.: Practical Byzantine fault tolerance and proactive recovery. ToCS **20**, 398–461 (2002). doi:10.1145/571637.571640

36. Merideth, M.G., Iyengar, A., Mikalsen, T., et al.: Thema: Byzantine-fault-tolerant middleware for web-service applications. In: Proceedings of IEEE Symposium on Reliable Distributed Systems, pp. 131–140 (2005)

37. Chun, B., Maniatis, P., Shenker, S.: Diverse replication for single-machine Byzantine-fault tolerance. In: USENIX Annual Technical Conference, pp. 287–292 (2008)

38. Kotla, R., Dahlin, M.: High throughput Byzantine fault tolerance. In: Proceedings of 2004 International Conference on Dependable System Networks, p. 575 (2004). doi:10.1109/DSN.2004.1311928

39. Nakamoto, S.: Bitcoin: a peer-to-peer electronic cash system, p. 9 (2008). doi:10.1007/s10838-008-9062-0, www.bitcoin.org

40. Pritchett, D.: Base: an acid alternative. Queue **6**, 48–55 (2008)

41. Gusella, R., Zatti, S.: An election algorithm for a distributed clock synchronization program (1985)

Improving Performances of an Embedded Relational Database Management System with a Hybrid CPU/GPU Processing Engine

Samuel Cremer[1,2]([✉]), Michel Bagein[2], Saïd Mahmoudi[2],
and Pierre Manneback[2]

[1] Computer Engineering Department, Haute Ecole en Hainaut,
Avenue Maistriau 8A, 7000 Mons, Belgium
samuel.cremer@heh.be
[2] Computer Science Department, University of Mons,
Rue de Houdain 9, 7000 Mons, Belgium
{michel.bagein,said.mahmoudi,pierre.manneback}@umons.ac.be

Abstract. End-user systems are increasingly impacted by the exponential growth of data volumes and their processing. Moreover, post-processing operations, essentially dedicated to ergonomic features, require more and more resources. Improving overall performances of embedded relational database management systems (RDBMS) can contribute to deliver better responsiveness of end-user systems while increasing the energy efficiency. In this paper, it is proposed to upgrade SQLite, the most-spreaded embedded RDBMS, with a hybrid CPU/GPU processing engine combined with appropriate data management. With the proposed solution, named CuDB, massively parallel processing is combined with strategic data placement, closer to computing units. Experimental results revealed, in all cases, better performances and power efficiency compared to SQLite with an in-memory database.

Keywords: In-memory database systems · Embedded databases · Relational database management systems · GPU

1 Introduction

Current systems have to deal with an exponential growth of data volumes they have to store, process and distribute. In recent years, numerous data management paradigms have appeared. Major improvements have been accomplished, especially in the area of Big Data systems and numerous NoSQL solutions have emerged. The different issues involved in current data growth concern as well data centers as end-user applications. Even if the current trend is in favor of lightweight applications, end-user systems must deal with more and more data. Whether with desktop platforms or mobile devices, numerous end-user applications embed an RDBMS (such as SQLite, MySQL embedded or MS SQL Server Compact). Such embedded RDBMSs usually serve as storage systems,

© Springer International Publishing AG 2017
C. Francalanci and M. Helfert (Eds.): DATA 2016, CCIS 737, pp. 160–177, 2017.
DOI: 10.1007/978-3-319-62911-7_9

as well as cache systems to reduce the number of interactions between clients and servers, and hence preserve the responsiveness of user interfaces. Low latencies of end-user applications are regularly difficult to maintain because, in current systems, many CPU cycles are dedicated to graphical user interface functionalities which consume more and more resources. Offloading the CPU from most of database processing's permits precisely to dedicate more resources for those greedy ergonomic features and hence to improve the reactivity of applications. Increasing client-side computing capacities enables the processing of larger data volumes and hence to reduce the number of client-server communications. Moreover, autonomy of mobile devices can potentially be increased by moving database operations towards GPU cores. Compared to CPUs, GPUs of smartphones are more efficient architectures while they are often under-exploited by non-multimedia applications.

In this paper a hybrid implementation over CPU and GPU is suggested in order to improve SQLite performances. It was decided to focus this research on SQLite because it is the most widely deployed database engine throughout the world[1] (it is part of the majority of smartphone operating systems, browsers, Dropbox clients, etc.). The processing engine of SQLite runs, like other RDBMS engines, in a purely sequential manner. Its performances can be improved by using all processing units of CPUs and GPUs. For numerous applications, GPU architectures are currently more efficient than CPUs [1] and have become essential in modern systems, even in small devices like smartphones. Table 1 shows that, compared to CPUs and for the same order of fee and power consumption, GPUs have overall higher amount of cores, better computing power and a higher memory bandwidth.

Table 1. CPU vs. GPU architecture.

	CPU	GPU
Reference	Core i7 6850K	GeForce GTX 1080
Number of cores	6 (12 threads)	2560
Frequency	3.6–4 GHz	1.733 GHz
Cache	15 MB (L3)	2 MB (L2)
Computing power (FP32)	768 Gflops	8,873 Gflops
Memory bandwidth	4×19.2 GB/s	298 GB/s
TDP	140 W	180 W

Performances of In-Memory DBs are closely related to available memory bandwidth. With their higher memory bandwidths, GPUs can potentially deliver better performances than CPU architectures for database processing. To benefit from this high bandwidth, global memory access latencies must be masked

---

[1] SQLite: Most Widely Deployed and Used Database Engine, http://www.sqlite.org/mostdeployed.html.

through a massive parallelism. This is one of the reasons why GPUs are designed for massive workloads and CPUs always outperform them for little jobs. The major contribution of this paper is to propose a data placement strategy allowing the exploitation of a clever parallelism offered by multicore CPU and GPU architectures. The benefit provided by the proposition here is the improvement of the responsiveness and the energy efficiency of applications.

The remainder of this paper is structured as follows: Sect. 2 presents the state of the art and the CuDB's position. Section 3 describes the internal architecture of the system, its processing and storage engines and how join queries are processed. Evaluation results are presented in Sect. 4, and this paper ends with outlooks and conclusion.

2   State of the Art

The idea of using hybrid CPU/GPU architectures to accelerate the data processing of relational database engines emerged in 2004 [2], some years before the arrival of GPUs with unified shaders and the release of general-purpose processing on GPU (GPGPU) frameworks. To the best of our knowledge, two main approaches have been proposed in the literature.

The first one emerged in 2007 with GPUQP [3]. This approach divides query plans into different action patterns which could be processed with different levels of parallelism, either on CPU or GPU platforms. The authors focused on the processing of single join-queries and contributed to provide base design architecture for most following works. So far, large majority of research in this field focused on very specific aspects of RDBMS, without providing a complete database engine. For example, OmniDB [4], is a system in which the authors paid more attention to the maintainability properties of source code. GPUDB [5], was created to demonstrate the potential performances of GPUs with the Star Schema Benchmark [6]. CoGaDB [7] which is mainly designed to study the generation of execution plans, and Ocelot [8], an extension of MonetDB, can also be mentioned. With Ocelot, the researchers have proposed a "kernel-adapter" approach to make a portable database engine across different hardware architectures. From what is known, the latest project which is closest to a DBMS is Galactica [9], but with a partial support of SQL, it is rather intended to be used in Big Data environments. With the exception of GPUTx [10], where the authors were focused on transaction management and their locking mechanisms, previous works were dedicated to online analytical processing (OLAP).

The second approach, initiated by Sphyraena [11], forces full query plan processing on the GPU side. With Sphyraena, single table queries can be processed by GPU cores with a single kernel call what produces a marginal overhead compared to GPUQP's principle (with GPUQP, each query requires multiple kernel launches). Given that a GPU-enabled replacement solution for SQLite must be fast with all sizes of datasets and also with simple queries; this concept seems to be more promising in terms of speed and efficiency improvements for embedded databases. The implementation of Sphyraena mainly suffers from numerous data exchange penalties through PCI Express bus, does not exploit CPU's parallelism, and a suboptimal join queries processing.

Most previous solutions are partial DBMS dedicated to OLAP, working with "read-only" databases. These different researches are more targeted to Big Data systems and do not encounter many of the issues of an embedded full relational database managements system. Most of previous solutions are not able to compute queries on little datasets faster than conventional systems, what makes them not suited to be embedded inside end-user applications. GPUs suffer from size limitation and the lack of extensibility of their memories what complexify processing of large datasets (>16 GB). OLAP databases are often significantly bigger than available GPU device memory what makes difficult to process them without performance degradation due to the PCI-Express bottleneck. The proposed solution, named CuD, targets to boost embedded RDBMS thanks GPUs, which is fully justified by the fact that embedded databases are generally smaller than available GPU memory. The aim is to improve the performances at the RDBMS engine level, which implicitly increases the responsiveness of applications, while leveraging capabilities of available hardware architectures. CuDB is a hybrid CPU/GPU fully "read-write" embedded RDBMS engine. The proposal targets a high performance solution for either personal computers (workstations and laptops), small devices (embedded systems) or even server clusters. The version of CuDB described in this paper was first briefly presented during DATA 2016 [12].

To the best of our knowledge, it can also be noted that only four commercial products, Kinetica [13] (previously known as GPUdb), MapD [14], Sqream DB [15] and BlazingDB [16], are database engines accelerated by GPUs. To be complete, note also there is a fifth system, Parstream, a precursor in GPU accelerated databases. Parstream is a big data oriented system where multiple GPUs have as task to manage the indexes of very large datasets. Parstream has been acquired recently by Cisco Systems [17]. Cisco Systems aims to provide a solution for analyzing high IoT data flows. Current commercial products are mainly oriented for geographic information systems (GIS) and OLAP what is far to meet the same constraints as an embedded RDBMS.

3   Design of Proposed System

3.1   Internal Architecture

Before understanding how the solution works, the architecture of SQLite will be presented briefly. As shown by Fig. 1, SQLite can be subdivided into four logical units. The first unit is the interface where the incoming SQL queries are received and results are sent back to user application. The second logical unit is the "SQL Command Processor" which produces query execution plans (opcode list). These opcode lists are comparable to an assembly style instruction list and are interpreted by the third unit of SQLite named "Virtual Database Engine" (VDE). This virtual machine is in charge of executing those opcodes on data stored and managed into the last unit of SQLite, the "Storage Engine" (SE).

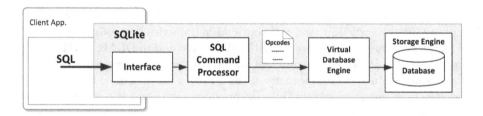

Fig. 1. Internal architecture of SQLite.

The properly subdivided architecture of SQLite makes it well-suited for an implementation of a new hybrid processing engine without changing anything at its API and SQL Command Processor.

With CuDB, the two first stages of SQLite are preserved in order to maintain SQL language support and to remain compatible with existing applications. VDE and SE units are the two components that intensely require the most resources. As shown by Fig. 2, CuDB embeds its own redesigned VDE and SE in order to exploit hybrid specificities of modern hardware architectures. The VDE of CuDB is designed as a Hybrid Virtual Machine (HVM) which incorporates two distinctive processing engines. One is dedicated to the GPU, and the other to the CPU, but both are based on the same parallel paradigm: each thread processes the query plan on its own data rows. By analogy with the SIMD (Single Instruction, Multiple Data) paradigm, the implemented approach follows what we called a SQPMD (Single Query Plan, Multiple Data) paradigm. The workflow through the two first stages does not differ from the SQLite implementation. Given that the query plans received by the HVM are initially intended to be processed by a single thread, they must be adapted in order to launch them for parallel processing. With the hybrid virtual machine, a unique produced query plan can be processed either by CPU or by GPU thread on a different dataset. It was deliberately chosen to not implement a simultaneous execution mechanism on CPU and GPU of a same execution plan for these reasons: (1) mainly avoiding the overhead due to synchronizations and data transfers between CPU and GPU,

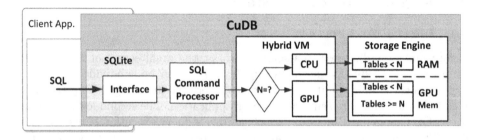

Fig. 2. Internal architecture of CuDB.

and (2) avoiding of using both architectures simultaneously to preserve some responsiveness of the graphical interface and other running applications.

To get some benefit from the high memory bandwidth between the graphic memory and GPU, the entire database is hosted directly by the GPU global memory (In-Memory-DB). The main advantage of this design is to prevent most of the data transfers between the CPU and GPU, avoiding unnecessary transfer latencies. This design is suited for most of end-user applications in which embedded databases are often smaller than available memory of GPUs. With the "in-memory" database and with the majority of the extraction queries, experiments have confirmed that performances depend primarily on the available memory bandwidth rather than computation power. As mentioned previously in Sect. 1, bandwidth between the GPU and its dedicated memory is often higher than that of CPU and its central RAM. This fully justifies the usage of GPU architectures for query processing, and especially when databases are able to fully reside in GPU device memory.

3.2 Processing Engine

A single data is processed faster by a CPU core than with a GPU core. This is one of the reasons why majority of GPGPU systems require a minimal amount of data for processing before becoming more efficient than CPUs. GPU systems must support the penalty of kernel launching, which is always accompanied by a data transfer through the PCI-Express. For small datasets this penalty produces a prohibitive overhead. This is why, in order to minimize the amount of kernel launches, the processing engine of CuDB is implemented following the SQPMD paradigm. Thanks to this paradigm, the efficiency threshold of the GPU processing engine gets lowered to only one thousand records. In order to deliver best performances independently of the size of datasets, a CPU processing engine is required for the processing of queries on tables of less than one thousand records. That is why, to make an efficient use of hybrid architectures, HVM chooses to execute processing, either on CPU or GPU resources according to the data volume they have to process.

With the GPU processing engine, once a result row is calculated by a thread, it is sent to the main memory through an asynchronous memory (pinned-memory). This technique allows threads to start processing next rows without waiting for the end of the transfers of the previous results, which increases the overall system performances. This transfer mechanism is obviously unnecessary with the CPU processing engine.

The implementation of the multicore CPU engine is based on the same design as the GPU version, but using POSIX threads instead of the CUDA framework. To ensure maximum performance with the CPU, several duplications of tables are also kept in the main memory. It is not necessary to duplicate the entire database in the central RAM memory, but only tables which are processed faster on CPU (tables with less than 1000 rows). With the first implementation of CuDB, a separate memory management was used, with RAM and GPU memory. The "unified memory" method provided by CUDA (from version 6) was tested to avoid keeping explicitly a permanent copy of some tables. With the "unified memory", CPUs and GPUs use a same pointer to access the data. This facilitates code writing with implicit management of memory transfers achieved by the driver. Experiments shows that severe slowdowns (between 2x to 9x) are introduced by the "overhead" of automatic memory management: the idea of using "unified memory" on the engine was abandoned. In the light of the high memory intensive workload of a database engine, the usage of unified memory did not permit to save memory space. The work presented in [18] also concludes that using "unified memory" can usually cause performance degradations.

3.3   Storage Engine

SQLite is one of the rare RDBMS which features a dynamic typing system for each value: this is called the "Affinity" mechanism. In order to preserve compatibility with existing applications, CuDB supports also dynamic data typing. However, such functionality entails a fairly high increase in complexity of treatments. Before reading a column value, the processing engine needs to read a typing header to know how to decode it. In case of a column value is used in a function or a predicate, a dynamic cast may also be required. Dynamic typing produces *de facto* a noticeable overhead which is strengthened by less consistent memory accesses due to custom size of each data. Performances of GPGPU solutions are very sensitive to coherency of memory accesses [19], which makes CuDB efficiency and performance suboptimal when it handles dynamically typed columns. As the majority of developers are mainly trained on statically typed database management systems like SQL Server, Oracle or MySQL, they are able to deal without "affinity" mechanism. This is why to reach the best performance; a selector for three different storage engine configurations is implemented. Like MySQL and MariaDB, each table of the database can use distinguished storage configurations. With these different storage engines, the database can be adjusted to its context while boosting the performances of applications with static data typing.

With CuDB, database insertions do not block the entire table and are processed asynchronously by the CPU. On each database update, data persistency is provided by a "write-only" mirror database saved on the hard drive. In the remainder of this subsection follows the description of the three storage engines.

Affinity Storage. This is the default storage configuration. It supports only dynamic typing similar to SQLite. Assumed that most of the records stored in a table do not occupy the same storage space, we have chosen to design this storage engine as a row-oriented structure in order to maximize data compactness of tables. A table is composed of two parts, table-header and records and a table-header is just a collection of records-pointers. As shown by Fig. 3, each tuple is always preceded by a header, which is required to support the "Affinity" mechanism. Depending upon the column number of the record, the record-header

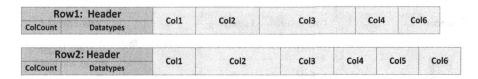

Row1: Header		Col1	Col2	Col3	Col4	Col6
ColCount	Datatypes					

Row2: Header		Col1	Col2	Col3	Col4	Col5	Col6
ColCount	Datatypes						

Fig. 3. Design of the record structure with "Affinity" storage configuration. On this example, Col5 of Row1 is not present because its value is Null.

does not always occupy the same number of bytes. To favor data compactness, Null and NaN values do not occupy memory space inside the record-value zone. They only consume four bits into the record-headers. This is the slowest storage configuration but, as counterpart, it promises to be compliant with the "Affinity" mechanism required by existing applications.

Strict Storage. Like the previously described storage setup, this configuration is a row-oriented data structure. The storage structure is similar to the "Affinity" engine configuration without the support of the "Affinity" mechanism. This fact implies that the table columns are statically typed like other database management systems. The suppression of the "Affinity" mechanisms allows the suppression of record-headers processing which procures a substantial performance boost. This storage mode is more intended for experimental purpose and is of minimal interest for the scope of this article.

Boost Storage. Like the "Strict" storage configuration, all data are statically typed without any support of the "Affinity" mechanism. The particularity of this storage configuration is that tables are stored as column-oriented data structures. As shown by Fig. 4, a table structure is split into three parts: (1) the table header which principally contains the column pointers, (2) the column-oriented data of the stored records, and (3) the variable length data in a row-oriented form.

Given that each CUDA thread works on its own record and on the same column, a column-oriented data structure is able to provide coalesced memory accesses which drastically reduce memory latencies (400 to 4 waiting cycles). Coalesced memory accesses can be obtained only with fixed-size columns. Fixed-size columns may pose some problems for variable-length strings and blob values. For those datatypes, we store only data pointers into the column-oriented storage zone. Those pointers point to associated values stored inside the row-oriented storage zone of the table structure to avoid waste of storage space. Fixed-length strings are entirely stored into the column-oriented storage region.

The main drawback of this storage engine is the higher data occupancy induced by fixed size columns. To avoid runtime errors with insertion queries, a datatype conversion mechanism can be activated. This mechanism converts values from the insertion queries to the datatypes imposed by table structure. As example, if a query tries to insert a string into an integer column, the string will be parsed and if a numerical value is found it will be converted and stored

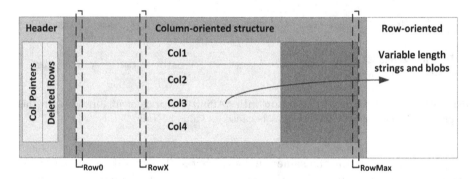

Fig. 4. Data structure of the "Boost" storage configuration. For each record there is a corresponding bit in the "Deleted Rows" bit vector that shows whether it is up to date or not.

otherwise the system stores a NULL value. This conversion mechanism can also be disabled to gain in performance with write intensive applications. With this storage mode, it is also possible to store single precision floats and fixed length strings which are not supported by SQLite which systematically uses double precision floats and variable length strings.

3.4   JOIN Queries

The SQL JOIN clause is one of the most used clauses with relational databases. With CuDB, the processing time of join queries highly depends on the selected storage engine. Note that the current version of CuDB does not yet support persistent indexes and consequently, what follows describe how joins with non-indexed columns are processed.

The query plan generated by SQLite for a join query with non-indexed columns, like in query (1), proposes the creation of a temporary indexation structure (B+ tree) for inserting records of table $t2$. For each record of $t1$, corresponding records of $t2$ are searched inside the transient indexed structure.

$$SELECT * FROM\ t1\ JOIN\ t2\ ON\ t1.col2 = t2.col3 \qquad (1)$$

This query plan has a complexity of $O(m.log(m))$ for creation and filling, plus $O(n.log(m))$ for parsing the data. Given that CuDB conserves the query compiler from SQLite, the generated query plans aim to be processed by a single thread. The challenge is to find the best way to automatically parallelize the proposed plan. To benefit from the massively parallel architecture of a GPU, a temporary indexation structure that can be filled concurrently by numerous threads is required. The classical B+ tree data structure used by the SQLite virtual machine is suboptimal regarding the GPU architecture because insertions cannot be accomplished concurrently.

Another constraint is that each thread makes its own search then; the structure exploration must be adapted to independent searches. Several existing GPU

B+tree structures were investigated, for example, T-trees and CSS-tree structures. However, it was found that these structures cannot efficiently be filled in a
parallel way what makes them not suitable for transient indexation with GPUs.
Moreover, with the GPU B++ tree [20], because each node of the tree counts
more than one thousand values, a sequential pass through is slower than with
a conventional binary tree. As temporary indexation structure, the approach is
to use a simple vector of records, which is filled in a parallel way by all the
GPU threads. As the order of insertions processed by GPU threads is *apriori*
unknown, it is assumed that the filled record-vector is unsorted. After filling, a
parallel sorting algorithm is launched in order to finalize the transient indexation
structure. In this way, a sorted vector is obtained where each thread can make
a dichotomic search in $O(log(n))$ operations.

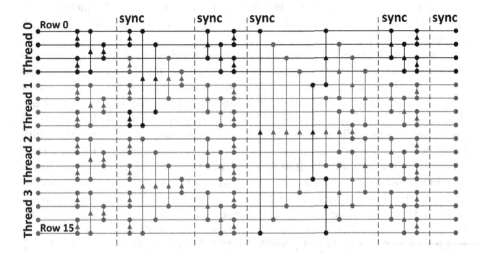

Fig. 5. Implementation of the bitonic sorter algorithm on GPU. Each thread sorts 4
rows, and thread synchronizations are required after every 4 comparisons. The bold
comparisons are processed by thread 0. Reprinted from [12].

To find an appropriate way to sort the record-vector, several GPU sorting
algorithms were investigated and one of the fastest algorithms was the radix
sort, with a time complexity of $O(n.w)$, in most cases, where n is the number of
keys of word length w. If all keys are distinct, w is at least equal to $log(n)$, but
the size of w can greatly increase with complex keys like strings. In a context
of database processing, the performances of a radix-sort can be very variable
depending on the key complexity, and the preference was to select an algorithm
which is independent of the key complexity. This is the main reason why a bitonic
sorting algorithm was implemented. Figure 5 shows the behaviour of such sorting
algorithm.

The bitonic sorter is a sorting network with a worst case complexity of
$O(n.log(n)^2)$. The complexity of the bitonic sorter is slightly worse than with a

radix sort, but the performances are stable regardless of the complexity of the keys. To reduce the overhead due to the number of synchronizations required by the bitonic sorter, some optimization techniques were carried out following [21]. Compared to most of other sorting algorithms, a bitonic sort does not require an additional data structure. Sorting is applied directly on the record-vector. Another significant advantage of bitonic sort is that it can be adapted to sort multiple columns, what makes it pertinent in a join mechanism implementation.

4   Evaluation

This section starts with a description of how the performances of CuDB where evaluated for selection queries. As current version of CuDB is mainly focused for delivering high performances for database reads, next subsections present speedups delivered for single table SELECT queries and JOIN queries. This section ends with a brief evaluation of the energy efficiency obtained with such a solution.

For the performance evaluations, the time required to process queries was measured with tables of varying sizes (between one hundred and one million rows). The tables consisted of four numerical columns followed by an 80 character string column. The columns were not indexed and the selectivity of queries was decreased starting from 10% for tables of one hundred records down to 0.1% for tables of one million records. This was motivated by the fact that with an embedded database and in an end-user application context, the amount of returned results is usually limited by the size of the user interface. The execution time of prepared statements was measured so that the compilation time of queries was not taken into account. The transfer times required to send the query plans to the GPU were considered, as well as the times needed by the GPU to send the results to the CPU. Different configurations of CuDB were compared with SQLite 3.8.10.2 and MySQL Embedded 5.7.11, both using in-memory databases. As performances can slightly fluctuate, each test was done a hundred times. The behaviour of the system was quite constant, and for better readability of this document, the average values are presented here. Due to limited resources, the experiments were run on a desktop with specifications shown by Table 2.

Given that CuDB is intended to be embedded inside end-user systems and not the majority of users dispose of a high end GPU, experiments were also conducted with an entry level GPU like the GT 740. The used system is a bit outdated but it is coherent and it can still deliver decent performances. In many areas, a GTX 770 can compete with a more up to date GTX 1060 GPU. To conform to an end-user context, GPUs where always installed as main graphic adapter and had to manage the graphical user-interface. This task causes a little overhead, which is mainly perceptible for queries on tiny datasets. Note that the maximal theoretical computing power of current CPUs is reached by using their SIMD instruction sets like SSE 4.2, AVX or AVX2. As example, to fully benefit from the power of AVX instructions, 8 additions and 8 multiplications must be processed concurrently within each CPU core. This is why, compared to GPUs, reaching the full potential of CPUs is often more complex.

Table 2. Hardware specifications.

	CPU	GPU1	GPU2
Reference	Core i7 2600K	GeFroce GTX 770	GeForce GT 740
Number of cores	4 (8 threads)	1536	384
Frequency	3.4–3.8 GHz	1 GHz	1 GHz
Cache (L2)	8 MB (L3)	512 KB (L2)	256 KB
Computing power (FP32)	243 Gflops	3,213 Gflops	762 Gflops
Memory bandwidth	2×10.6 GB/s	220 GB/s	80 GB/s
TDP	95 W	230 W	64 W

4.1   SELECT WHERE Queries

The different queries of this evaluation were applied to non-indexed tables with
columns of various data types (numerical and strings). The search conditions
used with the tested selection queries were principally comparison operators and
substring searches. Figure 6 shows the average speedups obtained by the different
test configurations with a standard implementation of SQLite as reference engine
with an in-memory database. The performances of the three storage engines were
measured, but in order to not overload this paper, only the slowest "Affinity"
and the fastest "Boost" storage mode are shown. With the hybrid engine of
CuDB, queries applied to tiny tables were processed by CPU cores and when
the table size exceeded one thousand records, CuDB switched from CPU to GPU
processing engine. Note that the CPU engine is directly able to deliver speedups
compared to SQLite and is also faster than MySQL 5.7.

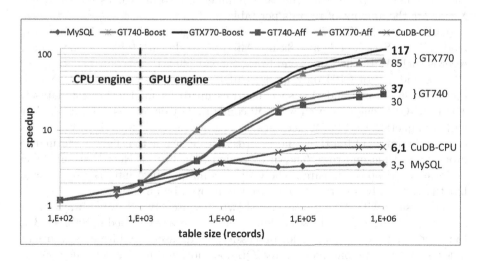

Fig. 6. Average speedups with SELECT queries.

For tables of one million records, and with the GTX770 GPU, speedups of 117x in Boost mode, 101x in Strict mode and 85x with the "Affinity" storage engine were obtained. As shown by the speedup of 6x delivered by the multi-threaded CPU processing engine, CuDB is also able to procure valuable speedups with systems that do not have a GPU. As the boosting of the client-side embedded RDBMS system was investigated, the system was evaluated with an entry level GPU. A "modest" GT740 procured already substantial speedups of 37x, 31x and 30x with the three different storage setups. Note that during the experiments, the highest speedups were obtained (411x with the GTX770, 107x with the GT740 and 15x with the CPU engine) for queries applied on fixed size string columns, and with a "WHERE col LIKE %susbstring%" search criterion. For the same query, MySQL 5.7 is 5,8 times faster than SQLite. These impressive speedups need to be put into perspective; since SQLite is not the fastest in-memory RDBMS when scanning single tables. The "memory" storage engine of MySQL 5.7 is in average more than 3 times faster than SQLite for large table scans. Compared to MySQL, CuDB running on a GTX770 is still 33 times faster.

4.2 SELECT JOIN Queries

In the previous subsection, it was shown that GPUs are very fast at processing full table scans thanks to their high memory bandwidth. In this subsection the results for JOIN queries will be presented. The tables were not indexed and the join conditions were applied on numerical data. The subset of join queries used for this evaluation includes "self-join" queries and join queries with two, three, four and five tables. All those queries are inner joins. The average results are shown in Fig. 7. Like the previous evaluations, and for better readability, the results of the "Strict" storage engine are not shown. The join queries were always done with tables with the same number of records. Values shown by the X-axis are the amount of records per table.

CuDB with a GTX770 GPU achieved average speedups of 44x in Boost mode, 21x in Strict mode and 8x in "Affinity" mode with tables of one million records each. With the same conditions, the GT740 GPU obtained speedups of 17x, 8x and 4x. In CPU only mode, CuDB is twice faster than SQLite with the smallest dataset and this speedup rises up to 4x for the biggest dataset. The CPU-GPU engine switch was still configured at one thousand records, but with join queries, the GT740 becomes faster than the CPU when involved tables count a minimum of ten thousand records. This explains the slight performance drop at five thousand rows (shown by region A in Fig. 7). Note that this little performance drop disappears if the GPU has no graphical interface to manage but this kind of hardware setup is rarely employed by an end-user.

The results with MySQL are not shown because MySQL was always much slower than SQLite with the set of join queries. As was explained in Subsect. 3.4, to reduce the time complexity for processing big datasets, SQLite uses transient indexes to compute join queries. MySQL does not, and implements such operations as nested loops. This results in multiple days of computing time for joining multiple tables of a million records, while CuDB needs less than a second.

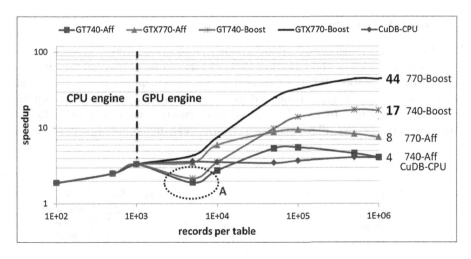

Fig. 7. Average speedups with JOIN queries.

Note that the peak speedups were obtained with "self-join" queries (66x with the GTX770, 28x with the GT740 and 4,4x with the CPU engine).

Unlike single table scans, the performance gaps between the different storage configurations were significant. For large datasets, switching from "Affinity" to "Strict" makes join queries more than 2 times faster, and switching from "Strict" to "Boost" produces again a 2x gain. As was explained in Subsect. 3.4, for each table join, the GPU has to perform a parallel sort. During the sorting operations, each GPU thread has to access and compare multiple tuples several times, unlike single table scans where each thread accesses only one record at a time. With "Strict" tables, GPU threads do not have to check each type of each data and with "Boost" tables, in addition to the static typing, the memory accesses are coalesced. That is one of the reasons why the performances of join queries are impacted by the choice of an appropriate storage engine. The other reason is that with the need to access multiple records inside a single thread, the sorting algorithm needs global synchronizations. With pre-Pascal GPU architectures (GTX770 and GT740 are Kepler GPUs) there is no robust way to implement global synchronization without using the CPU while maintaining a high degree of performances. With CuDB, most synchronizations imply a save and restore of the GPU execution context. This is also the reason why general speedups obtained with join queries are lower than with single table scans and that more data is required before the GPUs perform better than the CPU. New Pascal GPU architectures introduce a form of preemption mechanism what potentially could help to implement global device synchronization, but complementary investigations are required.

4.3   Energy Efficiency

Based on energy consumption measurements of all the queries with every platform configuration, for the following short report, an average of all energy efficiency ratios is proposed. Energy consumptions were acquired with an external
device and energy efficiency is defined as a ratio of energy consumed by SQLite
over energy consumed by tested platform. To ensure that heaviest queries do not
dominate these results, Fig. 8 shows the average calculated energy efficiencies.

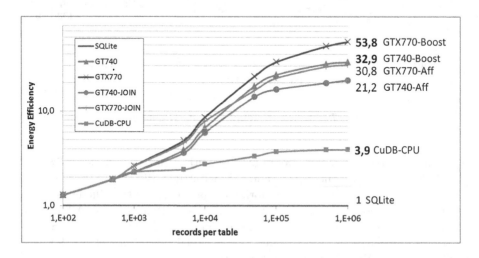

Fig. 8. Average energy efficiency (higher is better).

For a better readability, and like previous experiments, only results obtained
with CuDB in "Boost" and "Affinity" storage configuration are shown. During
the different experiments, and with the fastest storage engine configuration, the
GTX770 consumed nearly 54 times less energy than SQLite for the same job
with the biggest datasets. The much cheaper GT740 was also able to produce an
interesting energy saving of about 33 times and the CPU engine has consumed
nearly 4 times less energy than SQLite. Switching to the "Affinity" storage
engine configuration reduces overall performances what implies energy efficiency
degradation but the proposed system remains still more efficient than SQLite.

Faster memory accesses do not only impact performances but also the overall energy efficiency. Switching from the "Affinity" storage to the "Boost" configuration improves greatly memory accesses what has a significant impact on
energy consumption. These results confirm that the energy efficiency of embedded RDBMS can be significantly boosted by using a hybrid CPU/GPU query
processing engine.

5   Future Works

An important challenge is to overcome the limitations of the GPU memory capacity which is currently limited to 16GB for high end GPUs. The overhead of transient memory requirements involved in complex/nested join queries could also be larger than the physical GPU memory size. To overcome this size limitation, it is necessary to find an efficient mechanism to detect potential memory overflows, and optimal strategies to select, save and retrieve temporary data. The results of this investigation will also enable large database processing over multi-GPU.

A particularity of proposed solution is that the query plans generated by the SQLite Command Processor are preserved. Those query plans are designed to be processed on sequential CPUs and are not natively suited for parallel processing on GPU architecture, what potentially can slow down query processing. Redesign such query compiler could be an improvement of proposed system but it was deliberately chosen to conserve the SQLite query compiler for two main reasons: (1) to focus researches on parallel database processing engine and storage engines, and (2) to maintain a full compatibility with existing applications embedding SQLite.

Some implementation aspects of CuDB can be improved. For example, with the HVM, the CPU to GPU switching threshold is currently a static parameter. It could be interesting to design a self-calibration system that dynamically evaluates the best settings depending on the database features and system hardware specificities. Another improvement track is to improve the join engine. When the affinity mechanisms is turned off, and when the keys in the join conditions are only short values, switching from a bitonic sorter to a radix sorting algorithm could improve the general performances. When the join-queries are made on very small datasets, a trivial exhaustive search would be able to procure less overhead and improve the responsiveness of the engine. For these reasons, it should be interesting to add an indexation mechanism selector that, according to table sizes and data types, is able to switch to the most appropriated sort/search algorithm.

CuDB is still under development and as perspectives, the support of some additional clauses SQL and full indexation mechanisms will also be considered in order to be fully compliant with TPC-H and SSB benchmarks. When all previous improvements will be implemented, a port of the GPU processing engine on OpenCL can be considered to target other GPU manufacturers.

6   Conclusion

In this paper, CuDB, an embedded relational database engine that boosts the performance of SQLite by using multicore CPUs and GPUs, has been presented. To stay compatible with existing applications, CuDB preserves the SQLite API and the affinity mechanism can be enabled for existing applications. Weaknesses of GPGPU solutions for processing small amounts of data were also tackled by reducing the number of GPU kernel launches and by using a hybrid engine where lightest treatments remained on the CPU. It has been shown that GPU architectures can be exploited to speed up processing of RDBMS. Compared to SQLite with an in-memory database, peak speedups of more than 400x were achieved for substring searches on unindexed tables. The performances and the power efficiency of the presented solution were in all case better than SQLite. Energy measures have shown that faster data processing and memory accesses improvements reduce the overall consumption. The presented experiments have also confirmed that an entry level GPU is already able to provide noticeable accelerations.

References

1. Huang, S., Xiao, S., Feng, W.: On the energy efficiency of graphics processing units for scientific computing. In: IPDPS 2009, Sichaun (2009)
2. Govindaraju, N., Lloyd, B., Wang, W., Lin, M., Manochad, D.: Fast computation of database operations using graphics processors. In: SIGMOD/PODS 2004, Paris, pp. 215–216 (2004)
3. Fang, R., He, B., Lu, M., Yang, K., Govindaraju, N., Luo, Q., Sander, P.: GPUQP: query co-processing using graphics processors. In: SIGMOD/PODS 2007, Beijing, pp. 1061–1063 (2007)
4. Zhang, S., He, J., He, B., Lu, M.: Omnidb: towards portable and efficient query processing on parallel CPU/GPU architectures. VLDB Endow. 4(5), 1374–1377 (2013)
5. Yuan, Y., Lee, R., Zhang, X.: The Yin and Yang of processing data warehousing queries on GPU devices. VLDB Endow. 6(10), 817–828 (2013)
6. O'Neil, P., O'Neil, B., Chen, X.: Star Schema Benchmark (Revision 3, June 5, 2009). Technical report, UMass/Boston (2009)
7. Breß, S., Siegmund, N., Bellatreche, L., Saake, G.: An operator-stream-based scheduling engine for effective GPU coprocessing. In: Catania, B., Guerrini, G., Pokorný, J. (eds.) ADBIS 2013. LNCS, vol. 8133, pp. 288–301. Springer, Heidelberg (2013). doi:10.1007/978-3-642-40683-6_22
8. Heimel, M., Saecker, M., Pirk, H., Manegold, S., Markl, V.: Hardware-oblivious parallelism for in-memory column-stores. PVLDB 6(9), 709–720 (2013)
9. Yong, K., Karuppiah, E., Chong-Wee See, S.: Galactica: a GPU parallelized database accelerator. In: Third ASE International Conference on Big Data Science and Computing, Beijing (2014)
10. He, B.X., Yu, J.: High-throughput transaction executions on graphics processors. VLDB Endow. 8(5), 314–325 (2011)
11. Bakkum, P., Skadron, K.: Accelerating SQL database operations on a GPU with CUDA. In: 3rd Workshop on GPGPU, Pittsburgh, pp. 94–103 (2010)
12. Cremer, S., Bagein, M., Mahmoudi, S., Manneback, P.: Boosting an embedded relational database management system with graphics processing units. In: DATA 2016, Lisbon, pp. 170–175 (2016)
13. Kinetica: GPU-accelerated database for real-time analysis of large and streaming datasets. http://www.kinetica.com/
14. MapD: The World's Fastest Data Exploration Platform. http://www.mapd.com/
15. SQream DB. http://sqream.com/solutions/products/sqream-db/

16. BlazingDB: Blazing GPU Database. http://blazingdb.com/
17. Cisco has Completed the Acquisition of Parstream. https://lc.cx/orfA
18. Landaverde, R., Zhang, T., Coskun, A., Herbordt, M.: An investigation of unified memory access performance in CUDA. In: HPEC 2014, Waltham (2014)
19. van den Braak, G., Mersman, B., Corporaal, H.: Compiletime GPU memory access optimizations. In: ICSAMOS 2010, Samos (2010)
20. Kaczmarski, K.: Experimental B+-tree for GPU. In: ADBIS 2011, Vienna (2011)
21. Peters, H., Schulz-Hildebrandt, O., Luttenberger, N.: Fast in-place sorting with CUDA based on Bitonic sort. In: Wyrzykowski, R., Dongarra, J., Karczewski, K., Wasniewski, J. (eds.) PPAM 2009. LNCS, vol. 6067, pp. 403–410. Springer, Heidelberg (2010). doi:10.1007/978-3-642-14390-8_42

Author Index

Printed in the United States
By Bookmasters